SHEEP BREEDS
of New Zealand

SHEEP BREEDS
of New Zealand

Graham Meadows

REED

Published by Reed Books, a division of Reed Publishing (NZ) Ltd, 39 Rawene Rd, Birkenhead, Auckland. Associated companies, branches and representatives throughout the world.

This book is copyright. Except for the purpose of fair reviewing, no part of this publication may be reproduced or transmitted in any form or by any means, electronic or mechanical, including photocopying, recording, or any information storage and retrieval system, without permission in writing from the publisher. Infringers of copyright render themselves liable to prosecution.

ISBN 0 7900 0583 2

© 1997 Graham Meadows

The author asserts his moral rights in the work.
Cover and text design by Michele Stutton

First published 1997

Printed in New Zealand

contents

Author's note 6

part one

History 8
Aotearoa — Land of the Long White Cloud 14
Wool 20

part two

The Breeds 26

Useful Addresses 88
Glossary 89
Bibliography 91

author's note

While taking the photographs for this book I had the good fortune to travel to many parts of New Zealand, meeting and staying with the people who make up the backbone of rural communities. Without exception they made me welcome and were extremely helpful. I made many new friends and had some memorable experiences. One highlight was a morning spent on horseback high up in the tussock country of North Canterbury, photographing a sheep muster against the backdrop of the snow-covered Southern Alps.

Space does not permit me to list everyone I visited. May I just give my heartfelt thanks to the 55 families, partnerships and companies that allowed me to visit their stud farms and photograph their animals, and to the many others who provided advice, information, or photographic subjects.

Special thanks go to Paul Bowen, managing director of the New Zealand Agrodome, who was the prime instigator of this book; and to Wools of New Zealand for their support and the supply of statistics. The secretaries of the various sheep breed societies also provided information and assistance with the text: Tom Burrows (New Zealand Sheepbreeders' Association), Judith Pascoe (Cheviot Sheep Society of New Zealand, and Perendale Sheep Society), Chris Logan (Coopworth Sheep Society of New Zealand), John Morrison (New Zealand Wiltshire Sheep Breeders' Association), and Christine Ramsay (Romney Sheep Breeders' Association, and Southdown Sheep Society of New Zealand). I am indebted to the Rare Breeds Survival Trust, National Agricultural Centre, Kenilworth, England, for providing information on rare breeds and sheep breed development in Britain.

A book such as this is the product of a team effort. My thanks go to my editor, Alison Southby, who has once again been efficient and a pleasure to work with, and to all those involved with editing, proofreading and design.

Finally, any book requires a publisher, and I am gratified that Reed Publishing have shown their continued faith in me by agreeing to produce this one.

part one

history

Wild sheep
Domestication and development of breeds
The development of sheep breeds in Britain
The rise of the Merino

aotearoa — land of the long white cloud

The history of sheep in New Zealand
Sheep farming in New Zealand

wool

The characteristics and uses of wool
Wool shorn from living sheep
Wool from the skins of dead sheep or lambs
Converting wool into yarn and fabric

history

wild sheep

The ancient ancestors of sheep are known to have roamed Asia and Europe during the Pleistocene period (the Old Stone Age), which spanned from the end of the great Ice Age 250,000 years ago to the beginning of the Stone Age (9000 BC).

These animals were light-boned and agile to enable them to forage widely and escape predators, and they had hairy coats which were moulted in the spring. They gave rise to various types of wild sheep, of which four main types survive today: the Mouflon of Europe, Asia Minor and western Iran; the Urial of western Asia and Afghanistan; the Argali of central Asia; and the Bighorn sheep of northern Asia and North America. The Urial was probably the principal ancestor of today's domesticated sheep breeds, with the Argali having a smaller role in contributing to the breeds of India and the Far East. The Mouflon had some involvement in the process, but the Bighorn had none at all.

domestication and development of breeds

Domestication of the four common species of farm livestock (sheep, goats, pigs and cattle) took place during the Mesolithic period of the Stone Age (8000–6000 BC), at a time when humans were ceasing to be nomadic hunter-gatherers and were beginning to develop primitive agriculture and to live in permanent settlements. It commenced in south-west Asia, at about the same time in a number of different places, principally around three of the early centres of civilisation — the valley of the Tigris and Euphrates Rivers, the Nile valley, and the Indus valley — and then spread into Greece and other areas of south-east Europe. The pig was probably the first of these four species to be domesticated because like the dog, which had been domesticated 1–2,000 years earlier, it was omnivorous and could be fed on remnants of human food. Domestication of the goat and sheep came next, followed by that of cattle.

By the beginning of the Neolithic period (6000–3500 BC) domestic sheep were already smaller than their wild counterparts, and belonged to one single 'breed' which had short coarse wool mixed with kemps (coarse hairs). These early sheep were comparatively hairy, with far more hair fibres than wool fibres (a ratio of about 5:1). Males had heavy, three-edged spirally twisted horns, while females had short untwisted horns or were hornless.

Time Chart

250,000–9000 BC	Ancestors of sheep species present in Asia and Europe
8000–6000 BC	First domestication of sheep
1000 BC	Sheep breeding and wool processing in existence in Greece
100 BC–AD 476	Roman Empire refines processing of sheep's wool and milk
700–1000	Viking raiders introduce new sheep breeds to Britain
1000–1300	Cotswolds sheep industry emerges
1150–1200	Beni-Merines tribe introduces Merino to Spain
1500–1650	Portland breed emerges in Britain
1765	Exports of Merinos begin from Spain
1834	First breeding Merinos introduced to New Zealand
1882	Advent of refrigerated meat shipping from New Zealand
1910	New Zealand Romney most popular national breed
1950–1970	Development of Coopworth and Perendale breeds in New Zealand
1990s	Introduction to New Zealand of new breeds: the Oxford, Texel, Finnsheep, East Friesian, Awassi and Karakul

LEFT
Isolated on the St Kilda Islands of Soay and Herta, off the Outer Hebrides to the north-west of Scotland, the Soay is a primitive breed which is regarded as a link between wild and domesticated sheep. It sheds its fleece each year, and probably resembles the type of sheep kept by Neolithic farmers. It may be related to the wild Mouflon.

Horns

In most horned breeds males have larger horns than females, and in some breeds the females either have very short, rudimentary horns (scurs) or none at all. Horned breeds currently in New Zealand are the Dorset Horn and Drysdale, in which the horns are visible in both sexes; and the Awassi, Karakul and Merino, whose ewes either have scurs or no visible horns.

Sheep were kept and killed primarily for their meat, but they also provided a woollen skin, and ewes could be milked. As the Neolithic period progressed the proportion of hairs in the fleece was reduced and the quality of wool gradually improved, and the discovery that it could be spun, woven and felted to make cloth led to a widening of the domestication process. By 1000 BC the Greeks had developed a high level of wool processing and a system of conscious breeding that had created a number of sheep breeds that differed in appearance and use. Accumulated knowledge was gradually spread through Europe and Asia by trade and the migration of people.

The Scythians, who were originally a nomadic Indo-European people from an area to the north of the Black Sea, developed a fine wool, fragments of which were discovered by archaeologists in a tumulus grave in the Crimean peninsula and dated to about 500 BC. This find is interesting because the legend of the golden fleece is associated with the same area and the same period.

The Romans developed sheep breeding techniques even further, and also milked sheep, consuming the milk mainly in the form of cheese. They selected for white rather than coloured wool, and further improved its length, fineness and quality: samples of their wool show a hair to wool fibre ratio of 1:5. They eventually introduced their Roman Longwool sheep to all their provinces including Britain, which by the end of the Roman occupation was already an

ABOVE The Norfolk Horn was one of the primary producers of wool and milk in Britain during the Middle Ages, but no pure-bred Norfolk Horn survive today.

Tails

One interesting difference between wild and domesticated sheep that has not been fully explained is the length and width of the tail. Unlike their short-tailed ancestors, many domesticated breeds (particularly in Asia and the Mediterranean region) have long tails, some of them almost touching the ground. There are also fat-tailed breeds in Asia and Africa in which body fat is stored in the base and upper section of the tail.

important producer of cloth. After the fall of the Roman Empire in AD 476 many of the advances that had been made in the breeding of farm livestock were lost, and developments were the result of simple selection rather than conscious breeding.

ABOVE Adaptable and resilient, the Hebridean sheep is well suited to the sub-Arctic islands. A short-woolled breed, each animal grows up to three pairs of horns.

the development of sheep breeds in britain

Because the majority of sheep breeds in New Zealand originated from Britain, the origin and development of British sheep breeds is an integral part of the story of sheep in this country.

Between the seventh and tenth centuries the Vikings first raided and then settled in many northern and eastern areas of England and Scotland, and also the Isle of Man and Ireland. With them they brought sheep and cattle. In the north their sheep almost certainly influenced the development of local breeds, and the closest relatives to the original Viking sheep are probably the Hebridean and the Manx Loughtan breeds, both of which remained comparatively isolated on their respective islands and are now classified as Rare. They are short-woolled, short-tailed and their fleece is all of one colour, but their most remarkable feature is the ability to grow two, four or even six horns each. A typical mature ram can have four horns measuring 35–45 cm (14–18 inches) and weighing about 350 g (12 ounces) each.

The Roman Longwool sheep that had been left behind in Britain also played a role in the development of native breeds, being either assimilated into the indigenous (Soay-'type') sheep population or gradually developed into specialised populations in various parts of the country. These eventually gave rise to many of the earliest English breeds such as the Lincoln Longwool, Devon Longwool, Leicester, Wensleydale and Cotswold.

The Cotswold hills lie mainly in the county of Gloucestershire, and their name stems from two words: 'cotes', referring to the wattle enclosures that held sheep during the winter, and 'wolds', meaning hills; thus Cotswolds means 'the wolds of the sheep cotes'. By the Middle Ages (the tenth to the thirteenth centuries) Cotswold sheep had become famous for their wool, and the wealth from this wool trade built many of the beautiful churches and cathedrals that can still be seen in the area. In Britain's House of Commons the chancellor of the exchequer still sits on a sack of Cotswold wool (the Woolsack) as a symbol of the country's secure wealth.

ABOVE The Cotswold hills were England's first great wool-producing region.

Preventing the Extinction of a Breed

Throughout the Middle Ages Norfolk Horn ewes were kept for their wool and milk, the latter being made into cheese and sold in the cheese towns of Norwich and Ipswich. ('Wich' is an old English word meaning 'cheese town'.) From the seventeenth century on, the growing of root crops and a rotational system of farming were introduced, and this required sheep to be enclosed. The free-ranging Norfolk Horn eventually proved unsuitable for this, so between 1830 and 1850 it was crossed with the Southdown to develop a new breed, called the Suffolk. By 1919 only one flock of Norfolk Horns remained, and the last pure-bred animal died in 1973. Fortunately, in an attempt to save the breed during the 1950s, the two remaining Norfolk Horn rams had been crossed with Suffolk ewes. The female offspring were back-crossed to the Norfolk Horn rams, as were those of the next generation. Through further selective back-crossing the reconstructed population is now 80–90 percent pure and closely resembles the original breed.

Ironically the Cotswold breed is now rare, with only about 1,000 breeding females.

Crosses between Roman Longwool sheep and the native British (Soay-'type') sheep possibly resulted in a number of the tan-faced British breeds, such as the Welsh, Cheviot, Exmoor Horn and Norfolk Horn.

During the sixteenth century (Elizabethan times) the Ryeland breed was noted for the fineness of its wool, known as 'Lemster ore' (after the town of Leominster). It had a fibre diameter of about 26 microns, compared to the 21–24 of the Merino at that time (see page 13). There were, however, insufficient numbers of Ryeland sheep to provide any competition for the Merino, and most of the other British breeds of the time produced coarser wool of 28 microns or more.

Another breed to emerge during the sixteenth and seventeenth centuries was the Portland, probably one of the most direct descendants of the old tan-faced type of sheep that were among the earliest recorded in the British Isles. Some believe these may have originated from the Mediterranean, because some Portland ewes can lamb at any time of the year, normally giving birth to only one lamb after each pregnancy but lambing three or even four times in two years—a characteristic of many of the sheep breeds of the Mediterranean. Just to confuse the issue there is also an intriguing legend about Portland sheep arriving by swimming ashore from a sunken ship of the Spanish Armada. In the early 1800s the Portland was crossed with the Southdown to produce the Dorset Horn, a breed that can also lamb at any time of year, often having twins.

The turning point in sheep breeding came after the middle of the eighteenth century, when two Englishmen commenced the development of two British breeds. Robert Bakewell, an animal

BELOW
Some Portland ewes share with certain Mediterranean sheep the ability to lamb three times every two years, which has led to the theory that the progenitor of this breed inadvertently reached Britain with the Spanish Armada.

Products of breeding:
ABOVE
The Castlemilk Moorit was created in Scotland in the early 1900s through breeding of Manx Loughtan, Soay, Shetland and possibly other breeds.
ABOVE RIGHT
The Jacob sheep emerged during the eighteenth century and shares characteristics of both Viking and Mediterranean breeds.

geneticist, is credited with being the first person to introduce a systematic, 'conscious' breeding technique based on the premise 'like begets like'. He selected Leicester and Lincoln sheep for a more compact body and a greater amount of wool, and interbred them. He then interbred the offspring that also showed those characteristics, and continued to select and breed the offspring of subsequent generations in the same manner. In this way he developed the New or 'Dishley' Leicester (named after the village of Dishley), the ancestor of the Leicester Longwool which became known as the English Leicester.

To be successful this programme needed good records, and these became the basis for a flock book, a recording system that is still used today. The Leicester Longwool was later crossed with the Cheviot to produce the Border Leicester. Incidentally, Bakewell also applied his breeding techniques to cattle, developing a local type into the Dishley or Leicestershire Longhorn.

The other geneticist working in the same period was John Ellman of Glynde, Sussex, who developed the Southdown, which became the foundation breed for all the modern Down breeds. Other native sheep were improved by breeders using the same techniques, notably the Cheviot and Dorset Horn.

Cheese
These days the most famous cheese to be made from sheep's milk is Roquefort, from France. Other types are the Italian Pecorino and Ricotta Romana, the Sicilian Canelotti, the Portuguese Serra de Estrella, and the Greek Fetta.

Ornamental Sheep
Jacob Sheep
The Jacob is an ornamental breed of sheep, spotted black and white, that emerged during the eighteenth century. It was named from biblical references to Jacob (Genesis 30 and 31) who went to work for his father-in-law, Leban, and was paid for his labours in spotted sheep. By using spotted rams on all the best ewes he ended up owning most of the flock. Wool from these animals was almost certainly made into the 'coat of many colours' given by Jacob to his favourite son Joseph.

The origin of Jacob sheep is unclear. Piebald individuals occur in the Shetland breeds which are thought to be of Scandinavian origin, and the Jacob sheep may have two, four or even six horns, a characteristic of Viking sheep (see Hebridean, above). A popular theory is that they originated in the Mediterranean, where the local sheep have long tails and fine fleece, and where the Vikings are known to have taken their multi-horned sheep.

Castlemilk Moorit
This breed was created in the early 1900s at Castlemilk Park in Dumfrieshire, Scotland, by Sir Jock Buchanan-Jardine. A number of different breeds were used, including Manx Loughtan, Soay and Shetland. It has a characteristic short, tight, light-tan (moorit) wool which is popular for hand spinning.

the rise of the merino

Spain to Argentina 1569
Spain to Sweden 1723
Netherlands to the Cape 1780
Spain to England 1787
Cape to Australia 1797
Spain to USA 1802
Spain to Russia 1802
England to Australia 1805
Australia to NZ 1844
USA to Australia 1850

One of the most significant events in the history of sheep was the development of the Merino, a breed that is inextricably linked with the New Zealand sheep scene. The exact origins of the Merino are not known, although a popular theory is that it originated in North Africa around AD 750, and was developed by a nomadic Berber tribe, the Beni-Merines (after whom it is named), who took it with them during their movement into Spain during the second half of the twelfth century.

The unusually fine fleece of the Merino gradually came into demand in Europe, and to protect their trade the Spanish authorities banned all exports of sheep; the only exceptions were to some Spanish colonies which included Argentina, where Merinos arrived in 1569. By the late sixteenth century a substantial wool industry had been established in Spain, overseen by a government department called the Mesta.

A feature of the industry were the migrations, known as *transhumantes*, which followed specific routes and were organised by the Mesta. In late spring the sheep were driven in huge flocks to summer pastures in the northern uplands, then they were brought back to the southern plains of Estremadura and Andalusia for the winter. The sheep were gathered into flocks of about ten thousand, each flock guarded by 50 shepherds with their dogs, with the chief shepherd in front. They travelled up to 32 km (20 miles) daily and had right of pasturage over much of the kingdom, which caused ongoing complaints from the farmers and landowners along the way whose property suffered considerable damage.

During the seventeenth century opposition to the *transhumantes* increased and finally started to have an effect, with a subsequent decline in the power of the Mesta and the size of the Merino sheep population. Exports of Merino sheep commenced after 1765, and by 1802 there were more than 4 million animals in Germany alone. Concentrating on quality and sparing no expense, German breeders established the finest Merino sheep studs in the world. Other main centres of Merino breeding developed in Russia and America. From Holland Merinos were shipped to the Cape of Good Hope, from where they went to Australia and then New Zealand. Little more than a hundred years after their export from Spain they had become the most popular breed in the world.

ABOVE The spread of Merino sheep.
BELOW Guarded for centuries by Spanish herdsmen reluctant to share their bounty, the Merino is now the world's most populous breed of sheep, numbering some 3.3 million in New Zealand alone.

aotearoa – land of the long white cloud

ABOVE, LEFT
Ewes and lambs pictured in the Ruahine Ranges
CENTRE
Autumn colours set off this South Island flock.
RIGHT
Cormo ewes and lambs grazing in North Canterbury

New Zealand lies in the South Pacific between latitudes 35° and 47° S. The distance from the North Cape to the southern tip of Stewart Island is about 1,600 kilometres (1,000 miles). Superimposed on a map of Europe, it would stretch from northern Morocco to north-central France. Because of the differences in latitude within the country, the climate varies from cool temperate in the south to warm temperate or even subtropical in the far north. The other factors that influence the country's climate are its isolation as an island in the Pacific Ocean about 2,000 kilometres (1,200 miles) from Australia, the prevailing moisture-laden westerly winds, and its topography.

Running for much of the length of the South Island, the Southern Alps mountain range causes the incoming moist air to condense and fall as rain. As a result the west coast of the South Island is comparatively wet, with up to 6,000 mm (240 inches) of rain per year falling in Milford Sound, while the eastern areas are much drier. A similar pattern is found in the central North Island, where the high central ranges cause rain to fall in the more westerly areas, leaving the east coast (Hawke's Bay and Poverty Bay) comparatively dry. In the narrow neck of the north of the North Island there are no high mountains and rainfall is more evenly spread.

Many parts of the country receive regular rainfall throughout the year. With good soils and the addition of plenty of fertiliser, grass will grow all through the year in the north and for most of the year elsewhere.

LEFT
How the New Zealand sheep farm came into being.

the history of sheep in new zealand

The first sheep to be introduced into New Zealand were two Merinos, the survivors of a group of six brought from the Cape of Good Hope by Captain Cook on his second voyage in 1773. Landed in the Marlborough Sounds, they became ill and lived for only a few days. Later the early missionaries and whaling stations kept small flocks of sheep, primarily as a food source. For example, in 1814 the Reverend Samuel Marsden, a missionary with a keen interest in farming, introduced sheep into the Bay of Islands, but what breed they were and what became of them is unknown. It was 1834 before the first properly recorded breeding group, of Merino sheep, was imported to Mana Island in Wellington Harbour.

ABOVE Lincoln ewes and lamb

In the early days of European colonisation New Zealand's varied climatic conditions and environments dictated whether a breed was successful, and for the early settlers it was a question of trial and error. The environment in particular posed a challenge, for from a farming point of view it was virgin country. In some areas there were plains of tussock or native grasses, partly a legacy of earlier burning by the indigenous Maori, but much of the country was made up of steep hills covered in thick native forest. Large areas of this forest were felled, logged and burned by the settlers, who then sowed grass seed to establish pasture for their livestock. Areas treated in this way looked like battlefields, with tree stumps sticking out from the ashes and half-burnt vegetation littering the ground, and it was through this type of environment that the early sheep had to forage as grasses became established.

Soon after its introduction the Merino was found to be unsuitable in the wetter areas, being prone to footrot, so the settlers introduced hardy British breeds whose performance in their homeland's cool, wet conditions was already proven. The most successful

ABOVE
The Corriedale, one of the first two sheep breeds created in New Zealand.

of these were the Lincoln and English Leicester, excellent foragers whose strong coarse wool was less liable to snag and break on the numerous obstructions they encountered, and for half a century these two breeds were the mainstay of the New Zealand sheep industry. During the second half of the nineteenth century they were also cross-bred with the Merino to develop the first two New Zealand sheep breeds, the Corriedale and the Halfbred.

Many other British breeds were brought into the country. Some proved unsuitable, but the Southdown was successful in lowland areas. The Romney Marsh demonstrated an ability to thrive in most environments, especially those in the North Island with a high rainfall, and steadily gained favour.

Prior to 1882 the sheep industry was geared to the production of wool; farmers could not get fresh meat to overseas markets because of the long distances involved. But that year saw the advent of refrigerated shipping, which brought dramatic changes. Farmers started to change the make-up of their flocks to meet the British demand for mutton, and by the early 1900s the income from meat exports equalled that from wool. By then consumer preference had changed from mutton to lamb, and for lamb production the Border Leicester and the Shropshire had become the most popular sires, being cross-bred with Romney ewes.

By 1910 the much-improved Romney type, the New Zealand Romney, had overtaken the Lincoln and English Leicester to become the most popular breed in the country, a position that it maintains to this day. During the 1920s the Border Leicester and Shropshire were replaced as terminal sires by the Southdown, whose rams were crossed with Romney ewes to produce early-maturing prime fat lambs for export, a situation that lasted for about 40 years. The New Zealand Romney continued to gain popularity and by the early 1960s accounted for 75 percent of the national flock.

Between 1960 and 1990 the whole scene changed. Health-conscious Britons no longer wanted fat lamb, the Southdown went into a dramatic decline, and finally sheep meat prices slumped. Most commercial sheep farmers searched for alternative sires, mating Suffolk, South Suffolk, Dorset Down or South Dorset rams with their ewes to produce leaner lambs and improve wool production to meet an expanding wool market. The New Zealand Romney, though reduced in numbers, held its leading position, but two New Zealand breeds developed from it during the 1950s and 1960s, the Coopworth (Romney x Border Leicester) and Perendale (Romney x Cheviot), eventually became the second and third most numerous in the country. These dual-purpose breeds produced leaner lambs plus a readily saleable wool clip. The Borderdale, originally developed in New Zealand during the 1930s by crossing the Border Leicester and the Corriedale, also became popular.

The make-up of the national flock is still undergoing change. Romney numbers have been gradually falling, although it is still the most common breed, and new strains of Romney have also been developed.

New breeds were introduced during the 1990s, with the objective of using them for cross-breeding. The Oxford and Texel were brought in for their meat qualities, the Finnsheep for its fertility and resistance to facial eczema, and the East Friesian for its fertility and milk production.

ABOVE
A fine specimen of an Awassi ram
RIGHT
A flock of the 'improved' Romney: Up until the 1940s the breed was performing well under hill country conditions. From the 1950s, selection for wool quality became paramount and constitution secondary, and the Romney became less suited to hill country farming. Since 1970, two groups of breeders have been selecting for the Romney's original 'hill country' trait to produce the 'improved' Romney, and there are now at least three distinct strains.

Two other breeds were also introduced at this time: the Awassi and the Karakul, which belong to the group known as fat-tailed sheep, regarded as superior in many Middle-Eastern markets. It will be interesting to see what effect these breeds have on the future make-up of the national flock.

Creating a Breed
It takes many years and generations of inbreeding to 'fix' a new breed so that all the offspring are of a similar type. Therefore, for example, although the Coopworth was developed by crossing the Border Leicester with the Romney, a Romney ewe crossed with a Border Leicester ram produces a cross-bred, not a Coopworth lamb.

Sheep Breed Societies in New Zealand

The first formal breeders' organisation was the New Zealand Sheep Breeders' Association, which was founded on 28 May 1894. A code of rules was compiled and it was agreed that to be eligible for entry in the Flock Book the uninterrupted use of pure-bred sires was necessary since the year 1880; it also had to be verified that the flock was reputed to be pure-bred at the time.

The first Flock Book was published in 1895, and contained the histories of 291 flocks of the following 11 breeds: Lincoln (82); Border

ABOVE
Sheep and dogs form the archetypal New Zealand country scene.
BELOW
Huntaway at Rata Peak Station, South Canterbury

Leicester (65); Romney Marsh (51); English Leicester (48); Shropshire Down (18); Merino (14); Southdown (5); Cheviot (2); Cotswold (2); Hampshire Down (1); and Wensleydale (1). By 1898 the number of registered flocks had increased to 395 and all rams used in registered flocks were subject to individual identification and single entry.

A separate Romney Association was established in 1904. In 1924 the Corriedale Society became independent (rejoining the NZ Sheepbreeders' Association in 1988), and in 1926 a Southdown Society was established. The Cheviot Society was formed in 1959, followed by the Perendale Society in 1960.

Following the development of the Coopworth a separate breed society was formed in 1969. A Wiltshire Association was formed in 1986.

The New Zealand Sheepbreeders' Association is currently responsible for 25 of the 33 sheep breeds in this country: Borderdale, Border Leicester, Corriedale, Dorset Down and South Dorset, Dorset Horn, Drysdale, East Friesian, English Leicester, Finnsheep, Gotland Pelt, Halfbred, Hampshire, Lincoln, Merino and Poll Merino, Oxford, Poll Dorset, Polwarth, Ryeland, Shropshire, South Suffolk, Suffolk, Texel and White Headed Marsh. There are currently no societies for the Awassi and Karakul.

sheep farming in new zealand

Facts and Figures
New Zealand has about 20,000 farms carrying 50 million sheep, ranging from sea level to high up in the mountains at altitudes of 2,000 metres (6,600 feet) or more. With a human population of 3.5 million, New Zealand has a ratio of 14 sheep for every human, the highest in the world. Only Australia and Russia have greater numbers of sheep.

Wool

Annual wool production is 200,000 tonnes, an average of 5.4 kg (12 lb) of wool per head, which is high compared to many other countries. New Zealand exports more than 90 percent of its total wool production to 50 countries, mostly in the northern hemisphere. About 60 percent of the wool clip is ideally suited to carpet manufacture. Fine-woolled sheep such as the Merino are shorn once a year, strong-woolled breeds twice a year, and others every eight months (three times over two years).

Meat

Sixty-three million lambs are produced annually, of which about 24 million are exported, mostly in carcase form or as lamb cuts.

Sheep Farms

Sheep farms are of two main types: stud farms and commercial farms. Stud farms are specialised units set up to maintain or improve a specific breed, and produce pure-bred stud rams for sale to farmers running commercial flocks. Commercial farms keep ewes of one or more breeds or cross-breeds. The farmer usually splits these ewes into various groups (mobs) which are then mated by selected stud rams. These rams may be of the same breed as the ewes, thereby producing pure-bred replacements for the flock, or of different breeds, producing cross-bred lambs for either flock replacement or slaughter. A few farmers run both types of operation.

ABOVE
Cross-bred ewes like these East Friesian x Romney hoggets may become the mainstay of the New Zealand commercial sheep farm.
LEFT, RIGHT
The life cycle of the flock sets the yearly schedule for the farm, from the annual sheep draft to lambing.

wool

RIGHT
Cormo ram fleece showing its deeply crimped, white and bright quality

the characteristics and uses of wool

A number of factors influence the uses to which wools are put, and the prices they attract. Wool prices may be influenced by seasonal demand and supply; for example, when full-length fleece is dominating supply, price differentials for length become less obvious.

The important factors that dictate the price obtained for wool are:

Fineness (**fibre diameter, measured in microns**): influenced by the breed and time of shearing. The Chinese place great emphasis on the right micron value for cross-bred wool.

Length: influenced by the breed and the time of shearing.

Colour: evenness of colour and lack of staining are important.

Bulk: especially important in carpet manufacture.

Lustre: carpet manufacturers prefer less lustrous wools.

Vegetable matter: contamination with this reduces the price.

There is little demand for the coarser, lustrous types of wool with little fibre crimp, particularly when these wools become discoloured. Such wools have severe technical limitations in the production of machine-made carpets, and more particularly for use in knitwear. There is a steady move away from these wools to the less coarse cross-bred types (like Perendale), due to ever increasing demand from China for finer apparel wools, and the

drive by machine-made carpet manufacturers for greater automation which places tighter specifications on fibre properties.

Carpet wools

Carpet wools may be either second-shear or fleece wools. For second-shear wools a length of 75–100 mm commands a higher price than 50–75 mm. Fleece wools are longer and are generally processed on the semi-worsted system, for which a minimum length of 100 mm is preferred and in which freedom from vegetable matter is critical; the range is 0.1–0.2 percent, the lower level preferred.

Knitwear

Shorn hogget wools are used for knitwear: China dominates the buying of this type of wool.

Apparel

Apparel, or clothing, is made from the wool of woolly hoggets; these are hoggets that were unshorn as lambs. Perendales produce high-quality wool with fibres of 29–31 microns diameter and a staple length of 125–150 mm. This wool is sought after for worsted processing through to the highest quality apparel, and may be blended with Merino wools.

ABOVE
Woollen apparel, knitwear and sheep-skins are one of the prime bargains for tourists to New Zealand.
BELOW
The craft of hand-spinning wool is alive and well.

ABOVE
Trimming the sheep fleece after shearing
RIGHT
Shearing a Polwarth sheep: a professional shearer processes about 300 sheep a day.

Lambswool

All New Zealand cross-bred lambswool is naturally crimpy. The majority of these wools enter the 'Shetland' trade via woollen-spun yarn production where lack of bulk is of little concern. In addition the finished knitted or woven product undergoes considerable modification by milling techniques which mask the influence of fibre crimp differences. As a result fineness is more important than bulk, and good colour is of secondary importance. Staple length of 50–75 mm is most commonly specified.

Bedding products

Wool from second-shear lambs and short shears is used for these products. These wools are used as fillers in duvets and mattress overlays. Traditionally Down-type wools are used but others such as highest bulk Perendale are also suitable.

Wool is easy to dye compared with artificial fibres, but if stained yellow skirtings are included in wool they don't take the dye well, leading to inconsistent dye colour.

wool shorn from living sheep

Wool may be shorn from living sheep, or removed from their skins after death.

Wool Breeds

Wool from breeds that are specifically grown for wool production is classified as either fine, medium or strong. Fine wool has a fibre diameter within the range 15–24 microns. Strong wool has a fibre diameter of 30–41 microns and above.

Fine wool is commonly used in the manufacture of woollen or worsted fabrics for apparel (see below), whereas stronger wool is normally destined for carpet manufacture, upholstery and various coverings.

Meat Breeds

Wool from these breeds, from which the wool is a secondary product, is often (misleadingly) called cross-bred wool. It is commonly of medium strength (25–31 microns), and the finer wools are used in the manufacture of specialty garments and hosiery.

Shearing

Shearing is a skilled job that requires high levels of physical fitness and concentration.

ABOVE
Sorting and packing of wool is all-important for optimising the price gained at auction. The wool here is prepared for inspection.

Although farmers may shear a few of their own sheep, the bulk of the job is contracted out to shearing gangs. These shear the wool from the sheep and pack it into bales. Each shearer is paid according to the number of animals he or she has shorn. A good average is 300 a day, although in shearing competitions tallies of well over 400 in nine hours are regularly achieved. The current world record, set on 28 January 1997 by Darin Forde of New Zealand, is 720 ewes in a nine-hour day. The national competition, the Golden Shears, is held every March in Masterton.

wool from the skins of dead sheep or lambs

These may be used in their entirety, or treated to allow wool to be sliped (see below).

Woollen Skins
These are used for coats, rugs, etc. Karakul sheep are bred for the skin of their new-born lambs, which are killed within 24 hours of their birth. The resulting product is known as Persian Lamb or Astrakhan (the latter term is also used for a woven imitation). Kid gloves come from milk lambs that have not started to eat grass.

Wool Removed From the Skin (Slipe Wool)
Slipe wool represents about 13 percent of annual wool production. There are two methods of sliping — chemical and bacterial.

Chemical: The back of the skin is painted with a mixture of lime and sulphide, which works through the skin and weakens the roots of the wool, allowing it to be pulled off. This method produces a better skin but may damage the wool.

Bacterial: This method relies on the action of bacteria already in the skin. Skins are wetted and hung up in a warm room for several days. The bacteria multiply and rot away the roots of the wool fibres until they can be easily pulled off. This method produces excellent wool but a poor skin.

No longer considered as 'the black sheep' of the flock, coloured sheep produce wool of prime value for overseas textile markets, the carpet industry and domestic craftspeople. Selective breeding is more and more common in the attempt to create a desirable hue for a woollen fleece.
ABOVE
A self-coloured Merino ewe with a spotted lamb
LEFT
This coloured Perendale lamb sets off an attractive rural scene.

converting wool into yarn and fabric

There are two methods of making yarn — the woollen method and the worsted method.

Woollen method: This is the older of the two methods, and was the sole method used until the middle of the nineteenth century when machinery for mechanical combing was invented. It uses the shorter wools, which are carded but not combed, and in the resulting woollen yarn the fibres lie in various directions. This yarn is manufactured by weaving or knitting the wool into a soft, hairy fabric that feels warm to the touch, but it does not have the strength of worsted wool.

Worsted method: This method uses the longer wools. Mechanical combing removes the shortest fibres, called noils, leaving the longer fibres which are then drawn so that as far as possible they lie parallel to each other. After weaving or knitting this produces a strong, non-hairy yarn from which worsted fabric is manufactured.

part two

the breeds

Awassi
Borderdale
Border Leicester
Cheviot
Coopworth
Cormo
Corriedale
Dorset Down & South Dorset
Dorset Horn
Drysdale
East Friesian
English Leicester
Finnsheep
Gotland Pelt
Halfbred
Hampshire
Karakul
Lincoln
Merino & Poll Merino
New Zealand Romney
Oxford
Perendale
Poll Dorset
Polwarth
Romney (see New Zealand Romney)
Ryeland
Shropshire
Southdown
South Suffolk
Suffolk
Texel
White Headed Marsh
Wiltshire Horn & Poll Wiltshire

Awassi

RECOGNITION

Rams have large curled horns; ewes have either small scurs or minor horn development, or are polled. Roman nose, face clear of wool. Long pendulous ears. Head and neck have fawn through brown colouration. Body wool coarse and white. Legs and feet clear of wool, but hocks and feet may be coloured. The tail is surrounded by a large deposit of fat.

ORIGIN AND HISTORY

The Awassi is the most widely distributed and scientifically documented of the world's fat-tailed sheep breeds, and is the predominant breed in the Middle East. Intensive breeding in Israel has developed an improved dairy strain with exceptionally high capacity for milk production. Awassi embryos were imported from Israel via Cyprus in 1991 and transferred into ewes held in quarantine at Flock House, near Palmerston North. Offspring were released from quarantine in 1995. The original base flock consisted of ten rams and nine ewes which were transferred to Mathews Station, Tikokino, Central Hawke's Bay. There an intensive breeding programme was commenced using both embryo transfer and artificial insemination in order to build up numbers rapidly. The breeding programme is still based in Central Hawke's Bay, in the east of the North Island.

LEFT
Awassi rams

ABOVE
Awassi ewe and lamb

BREED CHARACTERISTICS

Multi-purpose, used for milking, meat and carpet wool. In New Zealand it was introduced for the live sheep trade to the Middle East.

A large-framed, fat-tailed breed suited to drier conditions underfoot. The breed characteristics of pendulous ears and fat tail are dominant and appear in the offspring when the Awassi is crossed with traditional New Zealand breeds. The wool is extremely fast growing.

Bodyweight

Ewes: 60–80 kg (132–176 lb).
Rams: 80–120 kg (176–264 lb).

Meat

Very lean. Most fat is concentrated in the tail, which weighs 8–14 kg (18–31 lb).

Wool

Extremely fast growing.
The fleece is composed as follows:
Wool: 40%. Often referred to as an undercoat. Fairly fine, with non-medullated fibres of about 25 microns.
Hair: 45%. Often referred to as an outer coat. Long, very coarse hair, which has medullated fibres of about 60 microns.
Heterotype: 10%. Often included in the outer coat. Long, coarse, semi-medullated fibres (in which the medulla may be discontinuous, fragmented or scattered) which are intermediate between hair and wool.
Kemp: 5%. Relatively stiff and brittle medullated fibres that shed naturally from the skin follicle. They have a considerable range in length and diameter (mean about 60 microns, range 20–120 microns).
Staple length: 150–200 mm (6–8 inches).
Fleece weight: 4.5–5.5 kg (10–12 lb).
Uses: Traditionally used in Berber and Persian carpets. Ideal for woollen carpets resistant to matting and tear. Awassi 50/50 cross-bred wool is being blended for use by the carpet trade.

Breeding/Lambing

Ewes have high milking ability and good maternal instinct. Lambing 100–150 percent.

Borderdale

ABOVE
Borderdale ram hoggets

RECOGNITION
Polled. White open face. Black nostrils. Poll covered in wool. Fleece comparatively long. Black hooves.

ORIGIN AND HISTORY
Developed in New Zealand, beginning in the 1930s with the crossing of the Border Leicester and Corriedale breeds to produce first-cross animals better suited to the dry, more fertile areas of the South Island. From 1970 such cross-bred animals were used in a controlled interbreeding programme to 'fix' the desired characteristics. The breed was recognised and the New Zealand Borderdale Sheep Society officially formed in 1977. Strict, mandatory performance criteria, including an annual cull of 30 percent from the ewe flock and 20 percent from the ram flock up to two-tooth stage, ensure that only genetically superior animals are used in order to obtain the best possible lambing percentages, growth rates and fleece weights.

The Borderdale thrives on irrigated or light pastures and in rolling hill country. It is found mainly on the plains, downland and foothills of Canterbury.

RIGHT
Borderdale ram

BREED CHARACTERISTICS

A medium-large, hardy, long-woolled breed, with a comparatively low susceptibility to footrot. Good growth rate. Ewe lambs can be mated at under one year old. Easy lambing. A dual-purpose breed, with good milk production. Good performance as a straight breed. Ewes are often used for cross-breeding with terminal sires from meat breeds.

Bodyweight
Ewes: 55–70 kg (121–154 lb).
Rams: 73–95 kg (161–209 lb).

Meat
Large, lean prime lambs with long carcase and good fat-lamb hindquarters.

Wool
Long, medium-fine cross-bred type. Emphasis on high wool weight.
Fibre diameter: Ewes 33–36 microns; Rams 30–40 microns.
Staple length: 100–150 mm (4–6 inches).
Fleece weight: Range 4.5–6 kg (10–13 lb); Average 5.5 kg (12 lb).
Uses: Home spinning. Commercial hand-knitting yarns and heavyweight apparel.

Breeding/Lambing
120–160 percent.

Numbers
516,000.

Border Leicester

RECOGNITION
Polled. White face and ears, clear of wool. Pronounced Roman nose. Black nostrils. Fleece strong and lustrous. Legs clear of wool. Black hooves.

ORIGIN AND HISTORY
Developed in the eighteenth century by Robert Bakewell (see History, page 12) on the English–Scottish border by crossing the Cheviot with the 'Dishley' or English Leicester. It inherited its characteristic bare head and legs from the Cheviot, a hardy, comparatively small and short-woolled breed that was already common on the hill country on either side of the border, whereas the English Leicester had a large deep body and heavy wool. The breed was introduced to Northumberland in 1767 and soon became well established in northern

BELOW Border Leicester ewe

RIGHT
Border Leicester ram calling

England and southern Scotland and known for its high fertility. The type was fixed by 1800 when the breed was registered.

The Border Leicester first arrived in New Zealand in 1859. Used mainly to produce sires for crossing with other breeds, it contributed high fertility and good mothering qualities as well as producing heavier lambs. It was crossed with the Corriedale to produce the Borderdale (see page 28) and the Romney to develop the Coopworth (see page 35). It continues to be used to increase fertility in commercial Corriedale, Merino and Romney flocks. Stud flocks are found throughout New Zealand, although the majority are in Canterbury, Otago and Southland.

BREED CHARACTERISTICS

Large, long-legged, with a pronounced Roman nose. A comparatively prolific breeder. A dual-purpose breed. Principally used for cross-breeding to improve the fertility and performance of other breeds.

Bodyweight
Ewes: 55–65 kg (121–143 lb).
Rams: 70–85 kg (154–187 lb).

Wool
Long, strong and lustrous. Individual staples are easily separated and end in a small curl.
Fibre diameter: 37–40 microns.
Staple length: 150–200 mm (6–8 inches).
Fleece weight: Range 4.5–6 kg (10–13 lb); Average 5.5 kg (12 lb).
Uses: Upholstery. Hand-knotted and machine-made carpet yarns.

Breeding/Lambing
Very good mothering ability. 110–160 percent.

Numbers
110,000.

BELOW Border Leicester ewes

Cheviot

RECOGNITION
Polled. White face clear of wool. Roman nose. Black nostrils. No wool on poll. Fleece bulky and of low lustre. No wool on legs. Black hooves.

ORIGIN AND HISTORY
One of the oldest British breeds, named after the Cheviot Hills in the border country between England and Scotland. It evolved as a hardy breed that would produce meat and wool off cold, wet, marginal hill country — characteristics which it still retains.

The Cheviot was first introduced into New Zealand in 1845, and a number of large flocks were eventually established in Southland and Otago in the South Island, and between Napier and Taupo in the North Island. However, the breed was not used extensively because better sheep country was being farmed which suited the bigger English breeds. In 1940 hill country trials were undertaken to evaluate the Cheviot, which was found to compare very favourably with other breeds in this environment. Following the introduction of aerial seeding and fertiliser topdressing in the 1950s, which allowed rapid development of newly cleared hill country, farmers discovered a need for a particular type of hill-country sheep. Breeding trials were undertaken by Sir Geoffrey Peren at Massey University, using specially selected Cheviot stud rams and Romney stud ewes, and from these was evolved the Perendale (see page 70).

Until fairly recently the Cheviot was a horned breed, but now all individuals are polled. The breed is found throughout New Zealand, mainly in ram-breeding flocks.

RIGHT
Cheviot ram

ABOVE
Cheviot ewe with one of her two lambs

BREED CHARACTERISTICS
Compact, short-legged. A good forager. Dual purpose. Rams are used as sires to produce cross-bred ewes (especially Cheviot x Romney), and as terminal sires for the production of lamb.

Bodyweight
Ewes: 45–55 kg (99–121 lb).
Rams: 60–75 kg (132–165 lb).

Meat
Fine-grained.

Wool
Bulky, low lustre.
Fibre diameter: 28–33 microns.
Staple length: 75–100 mm (3–4 inches).
Fleece weight: Average 2–3 kg (4.4–6.6 lb).
Uses: Carpet manufacture. Hand-knitting yarns, knitwear and tweeds.

Breeding/Lambing
Easy lambing. 90–110 percent.

Numbers
98,000.

Coopworth

RECOGNITION
Polled. White face with slight Roman nose. Black nostrils. Wool on the poll. Fleece long, medium strong and semi-lustrous, wool crimped. Legs clear of wool. Black hooves.

ORIGIN AND HISTORY
Developed in New Zealand at Lincoln College, Canterbury, during the 1950s and 1960s from the Border Leicester and Romney breeds, and registered in 1969. During the 1970s the breed base was widened to include white-woolled, white-faced breeds other than the Romney. It is named after Professor Ian Coop, who initiated the performance recording and selection on which the Coopworth Society bases its strict breeding requirements. It is mandatory for registered breeders to select stock for performance data which emphasise physical soundness, ease of management, high fertility, good mothering and milking ability, rapid weight gains and high fleece weight and quality.

The Coopworth has been remarkably successful in New Zealand where it has become the second most popular breed; it is widespread throughout the country on lowlands and improved hill country. There have been extensive exports to Australia, and it has also been exported to Eastern Europe and the United States.

RIGHT
Coopworth ram

ABOVE
Coopworth ewes

BREED CHARACTERISTICS

Medium-large, moderately hardy. Easy management with little shepherding. A broad pelvis and narrow shoulders makes for easy lambing. Good mothering and milking ability. High per head and per hectare production. Excellent performance on lowlands and improved hill country.
Dual purpose, with equal emphasis on meat and wool.

Bodyweight
Ewes: 55–65 kg (121–143 lb).
Rams: 75–85 kg (165–187 lb).

Meat
Lean and tender.

Wool
Coarse and long, semi-lustrous and well crimped. Medium bulk. Good colour (whiteness) and spinning qualities.
Fibre diameter: 35–39 microns.
Staple length: 125–175 mm (5–7 inches).
Fleece weight: Range 4.5–6 kg (10–13 lb); Average 5.5 kg (12 lb).
Uses: Heavier clothing. Carpets.

Breeding/Lambing
110–160 percent.

Numbers
4.9 million.

Cormo

RECOGNITION
Polled. Silky-soft face. Pink or mottled nose. Wool on poll and cheeks. Fleece dense, with medium-fine wool. Wool on legs down to feet. Hooves light to dark.

ORIGIN AND HISTORY
Originally developed in Tasmania in 1960 by Ian Downey of 'Dungrove' through the crossing of Superfine Saxon Merinos and Corriedale rams. Now found in many countries including Argentina, where the largest flock of over 110,000 Cormo sheep is run on El Condor station.

In New Zealand breed development was commenced in 1967 by Tim Ensor, of Cheviot, North Canterbury, who was eager to develop a sheep which combined the wool quality and whiteness of the Merino with the carcase quality of the traditional Halfbred or Corriedale that was run in his area. He selected top Merino ewes from the 'Grays Hills' Poll Merino stud and mated them to fine Corriedale rams from the Lammermoor stud. The resulting progeny were selectively mated to Corriedale and Cormo rams to establish the Jedburgh flock.

The Cormo is currently found in Otago, Canterbury and Marlborough in the South Island, and to a limited extent in the North Island.

RIGHT Cormo ewes and lambs

BREED CHARACTERISTICS
Large frame. Open face. Good foot conformation to withstand damp conditions. Ewes have very good milking ability and mothering instinct.

Dual-purpose meat and wool. Rams used for mating with Corriedale or Halfbred ewes for flock improvement.

Bodyweight
Ewes: 60–80 kg (132–176 lb).
Rams: 80–105 kg (176–231 lb).

Meat
Carcase long and lean. Ram and wether lambs very suitable for taking through to heavy weights for speciality cuts.

Wool
Dense fleece of medium length, highly resistant to the wool discolouration caused by high rainfall. Wool medium-fine, deeply crimped, white and bright.
Fibre diameter: 22–28 microns.
Staple length: 90–110 mm (3.5–4.3 inches).
Fleece weight: Range 5.4–6.2 kg (12–14 lb); Average: 5.8 kg (13 lb).
Uses: High quality knitwear, speciality hand-knitting yarns.

Breeding/Lambing
110–140 percent.

Numbers
About 65,000.

Corriedale

ABOVE Corriedale mustering at Glenovis Station, North Canterbury

RECOGNITION
Polled. Soft white face. Dark nostrils. Wool on poll. Fleece thick, with medium-fine wool. Wool on legs down to feet. Hooves dark.

ORIGIN AND HISTORY
The first indigenous New Zealand breed, pioneered in the 1860s by James Little, manager of Corriedale Station in North Otago. He mated fine-woolled Merino ewes, a hardy breed which thrives on high country and dry grasses, with long-woolled English Leicester or Lincoln rams, breeds better adapted to wetter, lush, pastoral conditions. Some Romney and Border Leicester rams were also used in the initial stages. The objective was to produce a sheep suited to pasture midway between that suitable for the parent breeds. Subsequent interbreeding of the progeny finally fixed the breed that Little called the Corriedale. The first committee sat in 1902, and the first registrations were in 1906.

New Zealand Corriedale sheep have been exported to many countries, including Australia, China, South Africa and the United States, but their biggest impact has been in South America. The Corriedale is now the fourth most popular breed in New Zealand, and vies with the Merino as the most popular breed in the world.

The breed is found in the drier parts of New Zealand, where the annual rainfall is between 500 and 750 mm (20–30 inches). It is most common in the South Island, in Marlborough and the eastern areas of Canterbury and Otago, but is also farmed in drier parts of the North Island.

RIGHT
Corriedale ram
BELOW
Corriedale lamb

BREED CHARACTERISTICS

A flexible, medium-sized breed suited to drier environments. It has a comparatively long productive life of up to seven years.

A dual-purpose breed, with equal emphasis on meat and wool. Rams are used for crossing with Romney or Perendale flocks to increase their body size, and to improve the fineness, weight, handling and colour of their wool.

Bodyweight
Ewes: 65–80 kg (143–176 lb).
Rams: 85–105 kg (187–231 lb).

Meat
Good length of carcase and muscling provides lean lambs for slaughter at an early age, or at a later age for heavyweight lamb grades.

Wool
Medium to fine, long-stapled, with a well-defined lock and pronounced and even crimp.
Fibre diameter: 28–33 microns (adults); 24–30 microns (hoggets).
Staple length: 75–125 mm (3–5 inches).
Fleece weight: Range 4.5–6.5 kg (10–14 lb); Average 5.5 kg (12 lb).
Uses: Adaptable to many uses, including medium-weight outer garments, worsteds and light tweeds, and hand-knitting yarn.

Breeding/Lambing
90–130 percent.

Numbers
About 2.8 million.

LEFT
Corriedales eating supplementary feed during winter

Dorset Down & South Dorset

ABOVE RIGHT
Dorset Down ewe and lambs
RIGHT
South Dorset ewe

RECOGNITION
Dorset Down: Dark brown face clear of wool. Dark brown legs covered with wool.
South Dorset: Brown face. Dark brown nostrils. Legs usually covered with wool.
Both types: Polled. Wool on poll. Fleece short, Down type. Black hooves.

ABOVE
Dorset Down rams

ORIGIN AND HISTORY

Developed in the middle of the eighteenth century in the county of Dorset in England by John Ellman, through the crossing the Southdown, Hampshire and local Dorset breeds. It was officially recognised in 1906. The breed was first brought to New Zealand in 1921, but soon died out. Further importations were made from 1947 onwards, and the breed quickly established a niche as a meat breed for the production of export lamb. The Dorset Down is suitable for a wide range of climatic conditions, and found throughout New Zealand from lowland pastures to hill country.

The South Dorset was developed in New Zealand during the 1950s through the crossing of Dorset Down rams with Southdown ewes. Registered in 1956 as the South Dorset Down, it was renamed the South Dorset in 1986. In 1997 a number of South Dorset breeders opted to use only registered Dorset Down rams as sires, and joined an appendix to the Dorset Down flock book. After three generations the offspring are eligible for full registration as Dorset Down stock.

RIGHT
A South Dorset flock

DORSET DOWN

BREED CHARACTERISTICS
Medium-large size. Rapid growth rate. Early maturity.
Meat breed. Rams are in demand as sires for terminal crossing with other breeds for the production of prime export lambs.

Bodyweight
Ewes: 65–80 kg (143–176 lb).
Rams: 100–130 kg (220–286 lb).

Meat
Carcase lean, high-yielding. Lean red meat.

Wool
Dense, Down type. Full-handling and springy.
Fibre diameter: 26–29 microns.
Staple length: 50–75 mm (2–3 inches).
Fleece weight: Range 2–3 kg (4.4–6.6 lb); Average 2.5 kg (5.5 lb).
Uses: Frequently blended with other wools to give extra elasticity and crispness. Also used in the production of high quality hosiery, fine knitting wools, bedding and furniture fillings, papermaking and felts.

Breeding/Lambing
110–140 percent.

Numbers
About 51,000.

SOUTH DORSET

BREED CHARACTERISTICS
Medium size. Rapidly maturing (second only to the Southdown).
A meat breed. Rams are used as sires for terminal crossing.

Bodyweight
Ewes: 70–85 kg (154–187 lb).
Rams: In excess of 100 kg (220 lb).

Meat
Large and meaty crossbred lamb carcase, well suited for the prime lamb export trade.

Wool
Bulky, fine, Down type.
Fibre diameter: 25–28 microns.
Staple length: 50–75 mm (2–3 inches).
Fleece weight: Range 2–3 kg (4.4–6.6 lb); Average 2.5 kg (5.5 lb).
Uses: Wool from slaughtered prime lambs is 'sliped' off the skins (by means of a chemical treatment) and comprises a large part of New Zealand's slipe wool production. It is used for hosiery and hand-knitting yarns.

Breeding/Lambing
120 percent or more.

Numbers
5,000.

Dorset Horn

RECOGNITION
Curled horns on both sexes, male horns larger than female. White face clear of wool, with noticeable pink skin. Pink nostrils. Wool on poll. Fleece short, Down type. Wool on legs to knees. Pale hooves.

ORIGIN AND HISTORY
Native to the county of Dorset in southern England, the breed was mentioned in literature in 1820 although it was not registered until 1891. It originated from the crossing of Portland sheep (now a very rare breed, see page 11) with Southdowns.

The breed first arrived in New Zealand in 1900 (some reports say 1897) but did not prove popular and little interest was shown in it until 1940, when ten ewes and a ram were imported from Australia. It has now largely been replaced by the Poll Dorset (see page 73), although it is found in small numbers throughout New Zealand.

BELOW
Dorset Horn ram

RIGHT
Dorset Horn ewe and lambs

BREED CHARACTERISTICS

Large. Able to breed twice a year, although most farmers either split the flock so as to produce spring lambs from one group and autumn lambs from the other, or else breed three times in two years. Hardy and vigorous, with rapid growth rate. Ewes show good mothering.
A meat breed. Rams used as sires for terminal crossing with other breeds.

Bodyweight
Ewes: 60–70 kg (132–154 lb).
Rams: 80–93 kg (176–205 lb).

Meat
Production of early and out-of-season lambs.

Wool
Very white, dense, and bulky Down type.
Fibre diameter: 27–32 microns.
Staple length: 75–100 mm (3–4 inches).
Fleece weight: Range 2–3 kg (4.4–6.6 lb); Average 2.5 kg (5.5 lb).
Uses: Hosiery, dress fabrics, flannels and fine tweeds. Dorset lamb skins are used in the fashion industry as linings for boots and shoes.

Breeding/Lambing
120–150 percent.

Numbers
300–400.

Drysdale

LEFT
Drysdale ram

RECOGNITION
Rams have heavy, curled horns; ewes have shorter, slightly curled horns. White face clear of wool. Black nostrils. Wool on poll. Fleece long, heavily medullated. Legs usually clear of wool. Black hooves.

ORIGIN AND HISTORY
In 1929 Dr F.W. Dry, a geneticist and hair biologist at Massey University, commenced research into the inheritance of hairiness in wool. In 1931 he was given a Romney ram that had a high abundance of hairy fibres (halo hairs), and mated it with Romney and

RIGHT
The Drysdale is one of New Zealand sheep breeding's success stories.
BELOW
A Drysdale flock changing pasture

Cheviot-cross ewes of the same wool type. By 1940 continued breeding trials had resulted in the isolation of a recessive gene (the Nd gene) which in sheep homozygous for the gene (Nd/Nd) produced horn growth and a fleece with large, medullated primary fibres. The Drysdale was developed during the 1960s into a commercial breed, and its medullated wool fibre is now a vital component in New Zealand carpet blends. It is found in almost all environments throughout New Zealand.

BREED CHARACTERISITICS
Medium-large. Very long, coarse wool. Shorn twice a year.
A dual-purpose breed, with equal emphasis on hard-fibre wool and meat.

Bodyweight
Ewes: 50–60 kg (110–132 lb).
Rams: 66–80 kg (145–176 lb).

Meat
Good quality lean meat.

Wool
Extremely long, strong and hairy. Heavily medullated.
Fibre diameter: 40+ microns.
Staple length: 200–300 mm (8–12 inches).
Fleece weight: Range 5–7 kg (11–15.5 lb); Average 6 kg (13 lb).
Uses: Carpet manufacture.

Breeding/Lambing
90–120 percent.

Numbers
600,000.

LEFT
Drysdale ewe and lambs

East Friesian

RIGHT
East Friesian ewe

RECOGNITION
Polled. White face clear of wool. Pink nostrils. Bulky, medium-coarse fleece. Thin, bare tail. Lower legs clear of wool. White hooves.

ORIGIN AND HISTORY
This long-established breed originated in the Friesland/Ost Friesland area in the north of Holland and Germany. In Europe it has been used either pure as a milking breed, as a crossing breed with other milking sheep breeds, or to improve fecundity and milk production in meat production breeds.

In December 1992 eleven pregnant ewes and four rams were imported into New Zealand from Sweden and entered a private quarantine station at Silverstream, near Dunedin. Thereafter a breeding programme was commenced using embryo transfer techniques, with only those animals derived from the embryo transfers eligible for release from quarantine. The first release occurred in March 1996, with 40 rams being sold while the remaining animals were held back to expand the flock numbers. However, there were substantial sales of semen, with an estimated 50,000 ewes of various breeds being artificially inseminated.

The first release of pure-bred ewes occurred in 1997. A New Zealand Sheep Milking Association was formed the same year, and the first commercial milking flocks with cross-bred East Friesians started production in 1997.

The base flock is maintained at Silverstream, but rams used for cross-breeding may be found on a number of farms throughout New Zealand, and many cross-bred flocks are becoming established. In 1997, about 100,000 ewes were mated to pure-bred East Friesians.

LEFT
East Friesian ram

BREED CHARACTERISTICS

A thin-tailed breed that has high fecundity and high milk production.

Particularly used for cross-breeding to improve the milk production and lambing percentage of other sheep breeds. The East Friesian will also become the basis for the establishment of a sheep milking industry for the production of fetta cheese and other sheep-milk products.

Bodyweight
Ewes: 75–95 kg (165–209 lb).
Rams: 100–125 kg (220–275 lb).

Meat
Very good lamb growth rates.
Carcase very lean.

Milk
The most productive sheep breed in the world, producing 500–600 litres per 210- to 230-day lactation.

Wool
Bulky, medium-coarse, white.
Fibre diameter: 35–37 microns.
Staple length: 120–160 mm (4.7–6.3 inches).
Fleece weight: Range 4–5 kg (8.8–11 lb); Average 4.5 kg (10 lb).
Uses: An ideal carpet wool.

Breeding/Lambing
Average of 280 percent in mature ewes.

Numbers
1,200.

English Leicester

RECOGNITION
Polled. White, wedge-shaped face. Black nostrils. Wool on poll. Fleece long, curly and lustrous. Wool on legs. Hooves black.

ORIGIN AND HISTORY
Developed in the county of Leicestershire, England, during the eighteenth century, when animal genetics pioneer Robert Bakewell began selecting the old type of Leicester sheep for a smaller, earlier maturing carcase, greater fat coverage and shorter legs. He used the Lincoln Longwool in his breeding programme and finally developed the 'Dishley' Leicester, a horned breed from which the Leicester Longwool, also known as the English Leicester, is descended. The Leicester Longwool was used in the development of other longwool breeds, including the Border Leicester, Cotswold and Wensleydale, but like many of the older breeds suffered a severe decline in popularity and in Britain is now classified as a Rare Breed.

The Leicester Longwool was first brought to New Zealand in 1843, and became known as the English Leicester. It was most popular in the wetter areas of the North Island where conditions were unsuitable for the Merino, which was prone to footrot. From 1880 to 1900 it was used in the development of the Corriedale (see page 38), and from 1890 onwards English Leicester rams became popular as crossing sires in the development of sheep best suited to wetter conditions and rough hill-country grazing. Among these was the Half-bred (see page 55).

RIGHT English Leicester ram

The breed is found mostly in Southland, Otago and Canterbury in the South Island; in the North Island the principal areas are the Wairarapa and Hawke's Bay. It is mainly found in ram-breeding flocks.

LEFT
English Leicester lamb calling

BREED CHARACTERISTICS
Large. Body deep and long. Hardy, with good fertility.
Dual purpose. Mainly used for creating cross-bred ewes and first-cross Half-bred (English Leicester x Merino) rams (see Halfbred, page 55).

Bodyweight
Ewes: 55–70 kg (121–154 lb).
Rams: 73–93 kg (161–205 lb).

Meat
Light in colour and of good texture.

Wool
Long and lustrous, of even length and fibre diameter. Good bulk. Locks curly, of medium width, showing a well-defined crimp.
Fibre diameter: 37–40 microns.
Staple length: 150–200 mm (6–8 inches).
Fleece weight: Range 5–6 kg (11–13 lb); Average 5.5 kg (12 lb).
Uses: Braids, suit linings, coatings, costume clothes and furnishing fabrics.

Breeding/Lambing
100–150 percent.

Numbers
15,000.

Finnsheep

RECOGNITION
Polled. White face. Pink nose. Very little wool on poll. Fleece white and lustrous. Long-bodied and long-legged. No wool on legs. White hooves.

ORIGIN AND HISTORY
The Finnsheep is an ancient breed, native to Finland, also known by the names of Finnish Landrace or Finn. It is one of several North European short-tailed Landrace breeds (see also Gotland Pelt, page 54).

The breed was introduced to New Zealand during the 1980s and released from quarantine in 1990. It is found in a wide range of climatic conditions throughout New Zealand.

RIGHT
Finnsheep ewes with Finn x East Friesian cross-bred lambs

BREED CHARACTERISTICS
Active. Highly fertile. Very good mothering ability. Highly resistant to facial eczema. Used in cross-breeding programmes with traditional sheep breeds to produce half-bred sheep with increased flock fertility and resistance to facial eczema. Being fine-woolled it may eventually become recognised as a useful dual-purpose breed.

Bodyweight
Ewes: 50–70 kg (110–154 lb).
Rams: 66–93 kg (145–205 lb).

Meat
Carcase lean and non-fatty.

Wool
Very white, lustrous wool. Fleece has good colour and bulk.
Fibre diameter: About 27 microns.
Staple length: 75–125 mm (3–5 inches).
Fleece weight: Range 2.5–4 kg (5.5–8.8 lb); Average 3.5 kg (7.7 lb).
Uses: Pure Finn wool is used for interior textiles.

Breeding/Lambing
175–250 percent.

Numbers
Under 5,000.

Gotland Pelt

RECOGNITION
Polled. Black face. Black nostrils. No wool on poll. Fleece grey and curly. No wool on legs. Black hooves.

ORIGIN AND HISTORY
The breed originated on the Swedish island of Gotland in the Baltic Sea. It is one of several North European Landrace breeds (see also Finnsheep, page 53). It was first imported into New Zealand in 1986, and released from quarantine in 1990. The breed occurs in only a few small collections in New Zealand; the major flock is owned by the Cornwall Park Trust and located in Cornwall Park, Auckland.

BREED CHARACTERISTICS
Lambs are born jet black, changing to the adult colour of grey by five months of age. A wool breed, developed for its high-quality, grey, curly pelt.

Bodyweight
Ewes: 50–65 kg (110–143 lb).
Rams: 66–86 kg (145–189 lb).

Wool
Grey curly wool of soft handle, high lustre and low bulk.

Fibre diameter: 27–33 microns.
Staple length: 150–200 mm (6–8 inches).
Fleece weight: Range 3–4 kg (6.6–8.8 lb); Average 3.5 kg (7.7 lb).
Uses: In Europe, lambs are slaughtered at the age of five months and their pelts are used for the manufacture of high fashion garments.

Breeding/Lambing
120–150 percent.

Numbers
About 300.

LEFT
Gotland Pelt have been present in New Zealand since 1986.

Halfbred

ABOVE
Halfbred ram

RECOGNITION
Polled. White face. Pink or black nostrils. Wool on poll and cheeks. Fleece thick and fine. Legs mostly covered with wool. White or black hooves.

ORIGIN AND HISTORY
Developed in New Zealand during the nineteenth century through the crossing of the Merino with one or other of the long-woolled breeds: English Leicester, Lincoln and Romney. The objective was to produce wool of a quality halfway between that of the contributing breeds. The term 'Colonial Halfbred' was used by the wool industry to describe this type of wool, and referred to the diameter of its fibre (micron count); it did not apply to the genetic mix of the animal that produced it.

Breeders quickly discovered that compared to the Merino the Halfbreds also had improved fertility and meat, and today they are farmed for their good performance in adverse climatic conditions. Registered Halfbred animals must be the progeny of registered Merino and

LEFT
Halfbred ewe and lamb

BELOW LEFT
A Halfbred flock at Mt Greba Station, North Canterbury

registered Longwool sheep of any pure breed, sheep of the first cross, or the progeny of registered Halfbred rams and registered Halfbred ewes.

They are found throughout the South Island foothills and in areas with light rainfall.

BREED CHARACTERISTICS
Medium size. Easy management. Dual purpose, with emphasis on wool production.

Bodyweight
Ewes: 40–55 kg (88–121 lb).
Rams: 53–73 kg (117–161 lb).

Meat
Medium-grained, tender, with no excess fat.

Wool
Fine to medium.
Fibre diameter: 25–31 microns.
Staple length: 75–110 mm (3–4.3 inches).
Fleece weight: Range 4–6 kg (8.8–13 lb); Average 5 kg (11 lb).
Uses: Worsted fabrics and fine knitwear.

Breeding/Lambing
85–130 percent.

Numbers
About 1.8 million.

RIGHT
Halfbred sheep at home in the South Island foothills

Hampshire

RECOGNITION

Polled. Black/dark-brown face free of wool. Black/dark-brown nostrils. Long, thick, black/dark-brown ears carried almost level and slightly curved. Wool on poll and cheeks. Fleece short, Down type. Black/brown legs, mostly free of wool. Black hooves.

ORIGIN AND HISTORY

Developed in the county of Hampshire, England, during the eighteenth century, probably through the crossing of Southdown rams with ewes of the old Wiltshire Horn and Berkshire Knot breeds, both of which are now extinct. It was fixed as a breed in 1889, and called the Hampshire Down. Although it arrived in New Zealand in 1861 it did not prove popular because it was a meat breed and at that time the major export market was for wool; the breed lapsed after 1901.

In 1951–2 two new flocks were imported from England, followed by stock from Australia. Now known as the New Zealand Hampshire, the breed still flourishes in all prime-lamb producing areas throughout New Zealand.

LEFT
Hampshire ewe

BREED CHARACTERISTICS

Large, alert and active. Tight, well-set shoulders; wide, flat loins. Wide, deep, well-developed hindquarters. Rapid growth rate with minimum fat.
A meat breed. Rams are used as terminal sires for crossing with many other breeds.

Bodyweight
Ewes: 60–75 kg (132–165 lb).
Rams: 100–120 kg (220–264 lb).

Meat
Excellent producer of prime lamb. Meat lean, sweet, and of good quality.

Wool
Fine. Down type. Free of black fibre.
Fibre diameter: 26–30 microns.
Staple length: 50–75 mm (2–3 inches).
Fleece weight: Range 2–3 kg (4.4–6.6 lb); Average 2.5 kg (5.5 lb).
Uses: Woollen hosiery, hand-knitting yarns and flannels.

Breeding/Lambing
110–150 percent.

Numbers
Under 4,000.

Karakul

RIGHT
Karakul ram

RECOGNITION
Horned and hornless animals occur in both sexes, although ewes are generally hornless. Rams' horns are long, spiral and outspreading; where present, ewes' horns are smaller. Head long and narrow with a Roman nose, covered with short, fine, lustrous hair. Black nostrils. Ears commonly long, thin and pendulous, but may vary in length through medium to short. Neck long and thin, chest narrow, body fairly long and deep. Hindquarters narrow with a drooping rump. Tail surrounded by a large deposit of fat. Long coarse fleece. At birth the coat colour of most lambs is coal black. The adult coat colour is commonly black or blue-grey, but may be various shades of brown or even white. Hooves black.

ORIGIN AND HISTORY
The Karakul may be the oldest breed of domesticated sheep, for archaeological evidence indicates the existence of Persian Lambskin as early as 1400 BC, and carvings of a distinct Karakul type have been found on ancient Babylonian temples.

The name comes from a village called Karakul in the former emirate of Bokhara, a region in West Turkestan to the east of the Caspian Sea. This high altitude region has low rainfall and scant desert vegetation, an environment which produced the hardiness and adaptability that remains in the sheep today. Although known as the 'fur' sheep, the Karakul was also a source of milk, meat, tallow and wool, the latter being felted into fabric or woven into carpeting.

Karakul sheep spread to the steppe region of southern Russia, the Middle East, Germany and other parts of the world. They were introduced into southern Africa in 1907, and now South

LEFT
Karakul lamb

Africa is one of the major exporters of Karakul pelts. Karakul sheep were first imported into New Zealand in the form of embryos from Zimbabwe, and implanted into surrogate mothers in April 1989. All subsequent offspring were released from quarantine in August 1994.

At the time of writing the breed is confined to one flock near Christchurch owned by Landcorp Farming Ltd.

BREED CHARACTERISTICS

A medium-sized fat-tailed breed that can withstand extreme feed shortages and exist in a harsh, semi-arid environment, but is sensitive to cold, wet conditions. Individuals are good foragers, walking considerable distances in search of food.
Farmed for meat production, especially live sheep for the Middle Eastern markets, and the production of lamb pelts (Astrakhan or Persian lamb) for garment manufacture.

Bodyweight
Ewes: 60–80 kg (132–176 lb).
Rams: 80–95 kg (176–209 lb).

Meat
Very lean. Most fat is concentrated in the tail.

Milk
In some European countries Karakul milk is used in the manufacture of butter and cheese.

Wool
The lamb pelt is light, silky and short-haired, with fibre in a range of curl patterns and colours. The adult fleece is composed of a long, thick, hairy outer coat with a fine woolly undercoat.
Fibre diameter: 37 microns.
Staple length: 200 mm (8 inches).
Fleece weight: Range 2–4 kg (4.4–8.8 lb); Average 2.8 kg (6 lb).
Uses: Lamb pelts are used for garment manufacture. The adult fleece produces superior carpet yarn.

Breeding/Lambing
Fairly long breeding season. Very good mothering ability with copious milk supply of high fat content (8.5 percent).
Lambing 80–110 percent.

Numbers
Less than 150.

Lincoln

RIGHT
Lincoln ram

RECOGNITION
Polled. White face. Black nostrils. Poll well covered in wool. Fleece thick, with long, broad, wavy locks. Wool on legs down to the feet. Black hooves.

ORIGIN AND HISTORY
Possibly the original longwool breed of England, the Lincoln Longwool originated in the Lincolnshire fens in eastern England and was recognised as an established breed about 1750. It was used by Robert Bakewell in his breeding programme to improve the old Leicester sheep and create the 'Dishley' Leicester (see English Leicester, page 51.) A few Lincolns were brought to New Zealand in 1840, but the first effective importation was in 1862. Hardy, and able to withstand cold, wet conditions, the breed was favoured by early settlers in the wetter regions of New Zealand, for it was able to forage on the regrowth emerging among the charred remains of the burnt-off native forest, and its coarse strong wool was more resistant than that of other breeds to snagging and breaking.

During the late 1800s the Lincoln was used in the development of the Corriedale (see page 38), and was crossed with Merino ewes to produce the popular Halfbred (see page 55). By 1900 there were 150 stud flocks, Lincoln rams were in demand as sires, and the breed was second only to the Merino in popularity. During the early twentieth century the Lincoln was gradually replaced by the Romney and the Corriedale, and since then demand for the breed has fluctuated. Progeny from the 30–40 studs that now remain are mainly used for crossing with Merinos to produce Halfbred rams for the South Island. The Lincoln is also crossed with other breeds such as the Romney.

The majority of studs are located in Canterbury, Otago and Southland.

BREED CHARACTERISTICS

Large, comparatively long-bodied and heavily built. A hardy breed, able to withstand cold, wet, harsh conditions.
Dual purpose. Mainly used for cross-breeding to give increased wool weights.

Bodyweight
Ewes: 55–70 kg (110–154 lb).
Rams: 73–93 kg (161–205 lb).

Meat
Lean carcase with long, meaty leg of mutton.

Wool
Long, coarse, strong and lustrous. A heavy fleece which opens freely with distinctive broad, flat, well-crimped, firm-handling locks.
Fibre diameter: 37–41+ microns.
Staple length: 175–200 mm (7–8 inches).
Fleece weight: Range 7–12 kg (14.5–26.5 lb); Average 8.5 kg (18.5 lb).
Uses: Products requiring high tensile strength, good lustre, and a soft handle. Specialised uses include upholstery yarns, hand-knitted carpet yarns, speciality knitting yarns, wigmaking and roller lapping. May be used as a substitute for or blended with mohair.

Breeding/Lambing
100–130 percent.

Numbers
10,000.

LEFT
Lincoln ewe and lamb

Merino & Poll Merino

ABOVE Merino rams at Moutere Station, Central Otago
ABOVE RIGHT Poll Merino ewe and lambs

RECOGNITION
Merino: Rams have long curled horns. Some ewes have very short horns (scurs).
Poll Merino: Polled.
Both types: White face. Broad pink nostrils. Wool on poll and cheeks. Rams and ewes have neck folds. Fleece very thick and bulky. Wool on legs. White hooves.

ORIGIN AND HISTORY
The oldest established breed in the world, and until recently the most numerous (see Corriedale, page 38). Thought to have originated in North Africa, and named after the nomadic Berber tribe, the Beni-Merines, who brought their sheep with them to Spain in the twelfth century. By the sixteenth century Merino wool was selling well in Europe and to protect this trade, exports of live sheep from Spain were prohibited. This law was changed in 1765 and thereafter Spanish Merinos were exported to many countries of the world, including France, which developed its own type, the Rambouillet French Merino. From French and Spanish stock large Merino flocks were eventually established in South Africa, North and South America, and Australia (1797), where the Booroola strain was later developed for its improved fertility.

Captain Cook introduced two Merinos into New Zealand in 1773, but the animals did not survive. The first breeding stock was introduced to Mana Island in Cook Strait in 1834, and additional animals were introduced from Australia during the 1840s. The breed was tried across the whole country, but it did not perform well in the wetter areas and was prone to footrot. Nevertheless, by the beginning of the twentieth century the Merino had become the dominant breed in New Zealand, with numbers reaching 14 million. But the development of refrigerated shipping in 1882 had already resulted in an increasing demand for meat breeds, and the Merino was gradually replaced on the lowlands and in the wetter areas by

ABOVE
Merino ram displaying the 'Flehmen' reaction when scenting a ewe in oestrus
RIGHT
Poll Merino ewes and lambs

dual-purpose breeds such as the Border Leicester, English Leicester, Lincoln and Romney, and by the New Zealand breeds derived from the Merino: the Halfbred and the Corriedale.

These days the Merino is mainly confined to the drier, mountainous and high plateau areas — the mountainous high country and lower hill country of the South Island, the Canterbury plains, and some areas of the North Island — but it still ranks as the fifth most popular breed in the country. The Booroola and Rambouillet strains of Merino are also present. Between flocks, wool characteristics range from medium to fine or superfine.

BREED CHARACTERISTICS

A fine-boned, medium-sized breed. Pronounc-ed neck and shoulder folds, especially in rams. Able to withstand drought, and a hardy and active forager.
A speciality fine-wool breed.

Bodyweight
Ewes: 40–55 kg (88–121 lb).
Rams: 53–73 kg (117–161 lb).

Meat
Fine-grained and very tender, with minimum fat.

Wool
Particularly fine, thick, bulky wool.
Fibre diameter: 19–24 microns.
Staple length: 65–100 mm (2.5–4 inches).
Fleece weight: Range 3–6 kg (6.6–13 lb); Average 3.5–5 kg (7.7–11 lb).
Uses: Quality woollen and worsted fabrics.

Breeding/Lambing
75–110 percent.

Numbers
About 3.3 million.

RIGHT
Merino lamb and ewe

New Zealand Romney

ABOVE
New Zealand Romney ewe hogget

RECOGNITION
Polled. White face. Black nostrils. Wool on poll and cheeks but allowing clear vision. Legs usually covered with wool. Hooves dark.

ORIGIN AND HISTORY
Developed in New Zealand from the Romney Marsh, a breed that evolved in the low wet marshlands called the Romney marshes in Kent, south-east England. For centuries the breed was effectively isolated from other parts of the country by the Kent Forest, and therefore remained comparatively pure while developing the strong constitution necessary to survive in an often harsh environment. These qualities have been passed on through the generations, and have been instrumental in its success.

The Romney Marsh was first introduced into New Zealand in 1853, and by the beginning of the twentieth century the breed had shown a remarkable growth in popularity. In 1904 a separate Romney Marsh breed society was registered, and the first flock book was published in 1905. Because of its adaptability to New Zealand conditions it very soon became the most popular breed in the country, a position it maintains today.

In 1956, in recognition of the changes it had undergone from its original ancestors, the name of the breed was changed to New Zealand Romney. The Romney was also involved in

RIGHT
New Zealand Romney ewe and lamb
BELOW
New Zealand Romney ram

the development of the Perendale (Romney x Cheviot) and the Coopworth (Romney x Border Leicester), and today is the predominant breed for crossing with terminal sires for meat production.

The Romney is well suited to high rainfall areas, and is widespread throughout the country in almost every type of farming environment.

BREED CHARACTERISTICS
Medium-large. Well-formed, heavy body. Good fertility and mothering ability, good lambing with little shepherding.
Dual purpose, with equal emphasis on meat and wool. Also mated to terminal sires for prime lamb production.

Bodyweight
Ewes: 45–65 kg (99–143 lb).
Rams: 60–86 kg (132–189 lb).

Meat
Large, lean lambs ranging on average from 15–22 kg (33–48 lb) carcase weight. Also hoggets for year-round production.

Wool
Heavy fleece with wool of medium lustre.
Fibre diameter: 33–40 microns.
Staple length: 125–175 mm (5–7 inches);
Second shear 50–100 mm (2–4 inches).
Fleece weight: Range 4.5–6 kg (10–13 lb);
Average 5.5 kg (12 lb).
Uses: About 60 percent is used in carpet blends. Other uses are for blankets, knitting yarns, overcoatings and furnishings.

Breeding/Lambing
90–150 percent.

Numbers
25.5 million.

LEFT
New Zealand Romney rams of the improved variety, at Wairere, near Masterton

Oxford

RECOGNITION
Polled. Brown face. Brown nostrils. Wool on poll and cheeks. Fleece short, Down type. Wool on legs. Black hooves.

ORIGIN AND HISTORY
An English breed, developed in the 1830s by crossing the Cotswold with a forerunner of the Hampshire, and using the resulting cross-breds to form the basis of the present-day breed. It first entered New Zealand in 1906, but generated little interest and died out. It was re-introduced in the 1980s, and was released from quarantine in 1990. The breed's capacity to produce a large, meaty carcase for further processing has stimulated interest from the meat industry, and it also grows the most wool of any of the terminal sire breeds.

There are a limited number of stud flocks in New Zealand (approximately 30), mainly in the South Island and the east of the North Island.

ABOVE Oxford ram
ABOVE RIGHT Oxford lamb

BREED CHARACTERISTICS
A large, heavy meat breed.
Rams used as terminal sires for prime lamb production.

Bodyweight
Ewes: 65–80 kg (143–176 lb).
Rams: 86–106 kg (190–234 lb).

Meat
Carcase large, with lean meat.

Wool
Down type. High bulk, with pigmented points.
Fibre diameter: 33–37 microns.
Staple length: 100–150 mm (4–6 inches).
Fleece weight: Range 3–4.5 kg (6.6–10 lb); Average 3.75 kg (8.2 lb).
Uses: Spinning.

Breeding/Lambing
90–120 percent.

Numbers
About 400.

Perendale

RIGHT
Perendale
ram hoggets

RECOGNITION
Polled. White face clear of wool. Slightly Roman nose. Black nostrils. Pricked ears. Wool on poll. Legs clear of wool. Black hooves.

ORIGIN AND HISTORY
A New Zealand breed developed at Massey University in the 1940s and 1950s, when it was decided to try to adapt the Romney to perform better on hill country with less palatable feed. Professor Peren developed the breed subsequently named after him by crossing selected Cheviot rams with Romney stud ewes, and then continuing to breed with the progeny to fix the type. It was first registered in 1960, and has become the third most popular breed in the country, as well as making its mark in Australia and elsewhere. The breed is found on grassland hill country throughout New Zealand.

RIGHT
Perendale ram
BELOW
Perendale ewe and her lambs

LEFT
Perendale ram

BREED CHARACTERISTICS

Of medium size with reasonable length of body. Active, hardy, and a good forager. Easy management, but requires quiet handling. Above average resistance to internal parasites. Few problems at lambing.
A dual-purpose breed, with equal emphasis on meat and wool.

Bodyweight
Ewes: 50–60 kg (110–132 lb).
Rams: 66–80 kg (145–176 lb).

Meat
Good lamb growth rates, even on hard hill country.

Wool
Long, of low lustre, and comparatively fine with free-opening staple. Medium-high bulk, with exceptional spring and resilience, and a high insulation factor.
Fibre diameter: 31–35 microns.
Staple length: 100–150 mm (4–6 inches).
Fleece weight: Range 3.5–5 kg (7.7–11 lb); Average 4.3 kg (9.5 lb).
Uses: Knitted garments, carpets, blankets.

Breeding/Lambing
100–140 percent.

Numbers
3.1 million.

Poll Dorset

RIGHT
Poll Dorset ewe and lamb

RECOGNITION
Polled. White face clear of wool, with noticeable pink skin. Pink nostrils. Wool on poll. Fleece fine and dense. Legs free of wool. Pale hooves.

ORIGIN AND HISTORY
Developed in Australia. In Victoria and New South Wales the Dorset Horn (see page 44) was the major breed, providing meat for the Melbourne and Sydney markets. However, the horns caused problems such as handling difficulties and damage to carcases. Selected Dorset Horns were crossed with the Ryeland and Corriedale breeds, and the resulting progeny (first-crosses) were mostly polled. Selected polled rams from this first cross were then mated back to Dorset Horn ewes, and only their polled progeny were kept for further breeding. Within a few years the Poll Dorset had evolved.

  In New Zealand, some breeders imported top Poll Dorset rams and mated them with Dorset Horn ewes, keeping only polled individuals for further breeding. Other breeders started their flocks using only pure Poll Dorset imports. The Poll Dorset has now largely replaced the Dorset Horn, and is found throughout New Zealand.

LEFT
Poll Dorset ram

BREED CHARACTERISTICS

Able to breed twice a year, although most farmers either split the flock so as to produce spring lambs from one group and autumn lambs from the other, or breed twice in three years. Hardy and vigorous, with rapid growth rate. Ewes show good mothering and excellent milking ability.

A meat breed. Rams are used as sires for terminal crossing with other breeds, and a limited number of first-cross ewes are used for breeding.

Bodyweight
Ewes: 60–70 kg (132–154 lb).
Rams: 80–93 kg (176–205 lb).

Meat
Production of early and out-of-season lambs.

Wool
Very white, denser than that from the Dorset Horn, and bulky.
Fibre diameter: 27–32 microns.
Staple length: 75–100 mm (3–4 inches).
Fleece weight: Range 2–3 kg (4.4–6.6 lb); Average 2.5 kg (5.5 lb).
Uses: Hosiery, dress fabrics, flannels and fine tweeds. Dorset skins are used in the fashion industry as linings for boots and shoes.

Breeding/Lambing
110–160 percent.

Numbers
About 23,000

Polwarth

RECOGNITION
Polled. White, soft face, clear of wool. Pink nostrils. Poll and cheeks covered with wool to eyes. Thick, crimpy wool on body and carried well down on belly and points. Wool on legs down to the feet. White hooves.

ORIGIN AND HISTORY
Named after the county of Polwarth in Victoria, Australia, where it was developed by 1880 through the crossing of the Merino and Lincoln breeds, the objective being to combine the fine wool of the Merino with the longer staple of the Lincoln. The breed was stabilised at three-quarters Merino and one-quarter Lincoln, thus producing a finer wool than the New Zealand-derived Merino crosses, the Corriedale and Halfbred. Polwarth sheep have been imported from Australia since 1932, and the present New Zealand flock contains the major bloodlines currently used in successful Australian studs.

Ideally suited to dry hill country, but also suitable for areas with higher rainfall because of the resistance of its wool to fleece rot. The Polwarth is mainly found in the Wairarapa, Marlborough, Canterbury and Otago.

BELOW
Polwarth ram

LEFT
Polwarth ewes at Matakanui Station, Central Otago

BREED CHARACTERISTICS

A medium-sized, fine-woolled, hardy breed. It is a dual-purpose breed, with major emphasis on wool production. The consistent basic micron count of the Polwarth produces a very dominant genetic wool influence when it is crossed with other breeds. Therefore rams are used for cross-breeding to 'fine down' flocks that have coarser wool, and to increase staple length and crimp, and improve wool quality in other breeds.

Bodyweight
Ewes: 50–60 kg (110–132 lb).
Rams: 66–80 kg (146–176 lb).

Meat
A good producer of prime lambs. Meat fine-grained, with very good ratio of lean muscle to fat.

Wool
High-yielding fleece with high resistance to wool rot. Wool of good commercial length, dense, even, soft-handling, bright and white in colour. Easy to scour.
Fibre diameter: 23–25 microns.
Staple length: 125–175 mm (5–7 inches).
Fleece weight: Range 4.5–6 kg (9.9–13 lb); Average 5 kg (11 lb).
Uses: Mainly for the production of fine knitting wools, woollen apparel and worsted cloth.

Breeding/Lambing
Can be bred at any time of the year; for example, where the climate permits in Australia they are lambed in autumn or spring. Some individuals have successfully lambed twice in one year. Easy, carefree lambing with very good mothering instinct. Lambing percentage 100–120 percent.

Numbers
About 160,000.

Ryeland

RECOGNITION
Polled. White face, wool on poll and cheeks. Pink/black mottled nostrils. Skin around eyes and nose dark. Legs usually covered with wool. Pale brown hooves.

ORIGIN AND HISTORY
An English breed developed more than 600 years ago by the monks of Leominster in the rye-growing areas of Herefordshire. Famous for its fine wool, known as 'Lemster ore'. Originally a small breed, it has been developed into a strong, hardy, medium-sized sheep that has been exported to most sheep countries throughout the world. Originally imported into New Zealand in 1903, rams were mainly used as sires in cross-bred and half-bred flocks to produce prime lambs. The breed's popularity peaked during the 1930s and 1940s, when there were several hundred stud flocks, but there are now fewer than ten. They are found in both the North and the South Island.

ABOVE Ryeland ewe and her lamb
ABOVE RIGHT Ryeland sheep, once famous for their wool

BREED CHARACTERISTICS
Medium-sized, stocky, with a broad straight back. Docile, easy to keep, and ideal for small farmers.
A meat breed. Rams are used as terminal sires for crossbreeding.

Bodyweight
Ewes: 55–60 kg (121–132 lb).
Rams: 73–80 kg (161–176 lb).

Wool
Short Down type. Comparatively fine, with dense staple, soft handle and good springiness. Practically no kemps or black and grey fibres.
Fibre diameter: 26–32 microns.
Staple length: 75–100 mm (3–4 inches).
Fleece weight: Range 3–4 kg (6.6–8.8 lb); Average 3.5 kg (7.7 lb).
Uses: Textiles (woven fabrics) requiring a smooth finish and good resilience; high quality tweeds and hosiery.

Breeding/Lambing
100–120 percent.

Numbers
About 500.

Shropshire

RECOGNITION
Polled. Soft black face and ears. Black nostrils. Wool on poll and cheeks. Fleece short, Down type. Black legs, the lower leg usually free of wool. Black hooves.

ORIGIN AND HISTORY
Originated in England in the mid-1800s, developed from several breeds popular in the West Midlands and Welsh border counties. Contributing breeds included the Longmynd (small, black-faced and horned), the Morfe Common (horned, speckle-faced and fine-woolled), the Southdown and possibly the Leicester. Very popular in the early part of the twentieth century, it is now classified as a Rare Breed in Britain, although numbers are once again increasing.

It first entered New Zealand in 1864. Numbers increased rapidly following the advent of refrigerated shipping (during the 1880s), for it was a breed that suited both the meat and wool trades because it produced a heavier fleece than other Down breeds. Rams were used for cross-breeding (particularly with Merino-type ewes) to produce lambs for export. In later years farmers in the Ruahine Range of the North Island, which is hard, high hill country, crossed the Shropshire with the Cheviot for prime lamb production.

There are currently about six stud flocks in New Zealand, all but one in the North Island, but there is rekindling interest in the Shropshire as a terminal sire. The breed does well in early dry country.

ABOVE
Shropshire ewe and lamb

BREED CHARACTERISTICS
Medium size.
A meat breed. Rams are used as terminal cross sires.

Bodyweight
Ewes: 55–60 kg (121–132 lb).
Rams: 73–80 kg (161–176 lb).

Wool
Short, Down type. Contains some kemp and grey or black fibres.
Fibre diameter: 26–30 microns.
Staple length: 50–75 mm (2–3 inches).
Fleece weight: Range 2–3 kg (4.4–6.6 lb); Average 2.5 kg (5.5 lb).
Uses: Woollen hosiery and knitting yarns.

Breeding/Lambing
100–120 percent.

Numbers
Less than 500.

Southdown

RIGHT
Southdown ewe

RECOGNITION
Polled. Mouse-coloured face with broad forehead. Dark nostrils. Ears and upper part of face covered with short wool. Short Down-type fleece. Wool on legs. Black hooves.

ORIGIN AND HISTORY
The oldest of the Down breeds, which evolved several centuries ago in the eastern stretch of the rolling chalk hills of Sussex, in the south of England. It was then a small, dark-faced breed with a short, light fleece, a long, narrow back and light forequarters, but with well-developed hindquarters. Scientific improvement was begun by John Ellman in the mid-1700s. Southdown sheep were later used in the development of the Dorset Down, Hampshire Down, Shropshire and Suffolk.

From the 1940s to the 1960s when consumer preference was for fat lamb, British buyers paid a premium for small, compact, early-maturing lambs, and the Southdown was a major player in this fat lamb trade. But following consumer reaction to the link between dietary fat and health problems such as heart disease, the breed underwent a dramatic decline between 1960 and 1980, and is now classified as a Rare Breed in Britain.

LEFT
Southdown ewe and triplets

The breed entered New Zealand during the 1840s (one source says 1860) and thereafter showed a steady increase. After the advent of refrigerated shipping in 1882 it became one of the favoured breeds for the production of export lamb. In the 1950s it was used in the development of the South Dorset (see page 41).

As in Britain, there was a dramatic decline in numbers during the period 1960–1980, although some breeders stuck to the breed and concentrated on breeding longer, leaner sheep. Others crossed Suffolk rams with their Southdown ewes to continue the development of the South Suffolk (see facing page). There are now signs of renewed interest in the leaner version of the Southdown, which is a much larger animal. The breed is suited to all classes of country throughout New Zealand.

BREED CHARACTERISTICS

Medium size. Long back, compact and well-muscled body, and comparatively short, sturdy legs. Earlier maturing and more heavily muscled than other Down breeds. Good fertility, with especially high lambing and survival rates when Southdown rams are crossed with ewes of dual-purpose breeds. Very high conversion rate of grass to protein, and economical to graze. Southdown sheep recover quickly from checks or setbacks.

A meat breed. Rams are used as terminal sires for cross-breeding with all dual-purpose breeds, including exotic crosses, to produce a fast-growing quality lamb carcase.

Bodyweight
Ewes: 50–80 kg (110–176 lb).
Rams: 100–150 kg (220–330 lb).

Meat
Southdown-cross lambs have a rapid growth rate, reaching carcase weights of 13–15 kg (28.6–33 lb) at 12–15 weeks. The meat is fine-grained, sweet and succulent, with an appealing bright-red colour.

Wool
Very dense, light, fine Down type with a distinct spiralled crimping which gives good bulk. Short staple.
Fibre diameter: 23–28 microns.
Staple length: 50–75 mm (2–3 inches).
Fleece weight: Range 2–2.5 kg (4.4–5.5 lb); Average 2.25 kg (5 lb).
Uses: Knitwear blends.

Breeding/Lambing
100–120 percent.

Numbers
Approximately 165,000.

South Suffolk

RECOGNITION
Polled. Dark brown face. Brown/black nostrils. Wool on poll. Short, Down-type fleece. Dark brown/black legs, free from wool. Black hooves.

ORIGIN AND HISTORY
A New Zealand breed developed during the 1930s by George Gould, a Southdown stud breeder. He combined the early maturing qualities of the Southdown (see page 79) with the leanness of the Suffolk to meet a demand for prime lamb with a minimum of fat, and then continued interbreeding to fix the South Suffolk's characteristics. Officially recognised as a breed in 1941, it has become increasingly popular. The breed is found on all types of country throughout New Zealand, both on stud farms and in commercial flocks.

BELOW
South Suffolk ewes at Mt Greba, North Canterbury

BREED CHARACTERISTICS
A large, comparatively heavy meat breed. Rams are used as terminal sires for crossbreeding for early prime lamb production.

Bodyweight
Ewes: 60–75 kg (132–165 lb).
Rams: 80–100 kg (176–220 lb).

Meat
High yielding carcase, ideal for further processing.

Wool
Short, fine, Down type.
Fibre diameter: 27–33 microns.
Staple length: 50–75 mm (2–3 inches).
Fleece weight: Range 2–3 kg (4.4–6.6 lb); Average 2.5 kg (5.5 lb).
Uses: Apparel and knitting yarns.

Breeding/Lambing
130–160 percent.

Numbers
94,000.

ABOVE LEFT
South Suffolk lambs
ABOVE
South Suffolk, a high-yielding meat breed

Suffolk

RECOGNITION
Polled. Black face clear of wool and covered in black glossy hair. Black nostrils. Short Down-type wool. Black legs, clear of wool. Black hooves.

ORIGIN AND HISTORY
Developed in the county of Suffolk, England, between 1830 and 1850 by crossing Southdown rams with black-faced Norfolk Horn ewes. It was fixed for type and named in 1859.

The Suffolk was brought into New Zealand in 1913 by George Gould, who later started the development of the South Suffolk (see page 81), and was followed by further imports from Australia and Britain. Today it is widespread throughout New Zealand, both on stud farms and in commercial flocks.

RIGHT
Suffolk ram

ABOVE
Suffolk ram
LEFT
Suffolk sheep at sunset, Mt Greba Station, North Canterbury

BREED CHARACTERISTICS

A large, hardy and robust breed, and a prolific breeder.
A meat breed. Rams are used as terminal sires for crossbreeding.

Bodyweight
Ewes: 60–80 kg (132–176 lb).
Rams: 80–106 kg (176–233 lb).

Meat
Lambs are fast-growing with a lean carcase, and ideally suited to the export trade in packaged lamb cuts.

Wool
Good quality. Short, fine, Down-type, but of comparatively low crimp and may contain dark hairs.
Fibre diameter: 30–35 microns.
Staple length: 75–100 mm (3–4 inches).
Fleece weight: Range 2.5–3 kg (5.5–6.6 lb); Average 2.7 kg (6 lb).
Uses: Hand-knitting yarns, tweeds, flannel and dress fabrics.

Breeding/Lambing
110–150 percent.

Numbers
About 55,000.

Texel

RECOGNITION
Polled. White face clear of wool. Black nostrils. White pricky ears. Low, wide body. Legs clear of wool. Dark hooves.

ORIGIN AND HISTORY
Developed on the island of Texel, off the coast of Holland, with records going back to the early 1800s. Outside crossing of the breed has been banned for over 120 years. The first imports to quarantine in New Zealand were in 1983 and 1984, and the first sheep were released in 1990. They are now making a significant contribution to the country's prime lamb trade and are found throughout New Zealand.

RIGHT
Texel were first released in New Zealand in 1990.

BREED CHARACTERISTICS
A meat breed. Rams are used as terminal sires for cross-breeding; also used to introduce a Texel component into ewe flocks.

Bodyweight
Ewes: 50–65 kg (110–143 lb).
Rams: 66–76 kg (145–167 lb).

Meat
Comparatively coarse-grained and later maturing than some other meat breeds. Renowned for its well-muscled, lean carcase with a high dressing-out percentage.

Wool
Speciality fleece of high bulk, medium coarseness and moderate length.
Fibre diameter: 33–37 microns.
Staple length: 75–125 mm (3–5 inches).
Fleece weight: Range 2.5–4 kg (5.5–8.8 lb); Average 3.25 kg (7 lb).
Uses: Futon trade. Hand-knitting yarn.

Breeding/Lambing
90–120 percent. Texel-cross lambs are hardy, with low losses from birth to weaning.

Numbers
Approximately 420,000 and growing rapidly as breeding-up flocks reach pure-bred status.

White Headed Marsh

RECOGNITION
Polled. White face. Black nostrils. Wool on poll and cheeks. Large body. Legs usually covered with wool. Black hooves.

ORIGIN AND HISTORY
Developed in the North Sea marshes of Germany during the middle of the nineteenth century, when North German Marsh sheep, the local milk sheep and various imported British longwool breeds (including the Cotswold) were interbred. It later became well established in northern Germany and Denmark. A hardy breed, it is capable of withstanding wet and very cold climatic conditions.

Imported into New Zealand under quarantine in 1986, it was released in 1990. It is run in stud flocks, and also used to inject size and higher lamb production into Perendale and Romney flocks. Evaluation is still continuing. At present there are a limited number of stud flocks in the South Island and the east of the North Island.

ABOVE LEFT White Headed Marsh ram
ABOVE White Headed Marsh ram hogget

BREED CHARACTERISTICS
Similar to the Romney but much larger and heavier. Performs very well in wet conditions. Dual purpose, with equal emphasis on meat and wool.

Bodyweight
Ewes: 70–80 kg (154–176 lb).
Rams: 93–106 kg (205–233 lb).

Wool
Similar to that of the Romney, but stronger and slightly coarser, with a higher percentage of medullated fibres giving better bulk. There are two distinct strains of the breed, one with finer wool of 31–38 microns (similar to the Perendale), the other with coarser wool of 40+ microns (similar to the Drysdale).
Fibre diameter: 31–44 microns.
Staple length: 125–175 mm (5–7 inches).
Fleece weight: Range 4.5–6.5 kg (9.9–14.3 lb); Average 5.5 kg (12 lb).
Uses: Principally for carpet manufacture.

Breeding/Lambing
130–150 percent.

Numbers
Under 500.

Wiltshire Horn & Poll Wiltshire

RECOGNITION
Horned or polled. White face clear of wool. Black nostrils. Large body with very light fleece. No wool on crutch and belly. Black hooves.

ORIGIN AND HISTORY
The British Wiltshire Horn is one of the oldest surviving sheep breeds. Its origin is obscure, but it resembles the ancestral sheep that existed in medieval times, from which arose the other white-faced, short-wool breeds found in the south of England. Until about the end of the eighteenth century the Wiltshire Horn was the predominant breed to be found on the Wiltshire Downs, after which it was named. It was used in the development of other Down breeds, such as the Oxford, Hampshire and Dorset. Like many breeds it eventually fell from favour, and became classified as a Rare Breed in Britain.

ABOVE Wiltshire Horn lamb
RIGHT Poll Wiltshire ewe and lamb with Wiltshire Horn lamb

The Poll Wiltshire was evolved in Australia by crossing Poll Dorset rams with Wiltshire Horn ewes, then back-crossing to the Wiltshire Horns. The Wiltshire Horn first entered New Zealand in 1972, and subsequently the poll gene was introduced. The majority of sheep in the country are of the poll variety.

The breed is widely distributed throughout New Zealand.

BREED CHARACTERISTICS
An unusual breed in that it has very little wool, which it sheds annually, so it has no dags and a very low susceptibility to fly strike. A strong foraging ability and long legs make it particularly well suited to roaming over wide areas, and it does well in arid climates and on poor pastures with little shade. These attributes have led to it being exported from Britain to many countries, particularly those with hot climates. In recent years exports from New Zealand have gone to the Pacific Islands, Asia, and North and South America. Primarily a meat breed.
Rams are used as terminal sires for cross-breeding.

Bodyweight
Ewes: 60–75 kg (132–165 lb).
Rams: 110–142 kg (242–312 lb).

Meat
The breed is selected for rapid growth to produce heavy, lean lambs suitable for further processing. It has a large lean carcase that is claimed to be superior to other breeds or crosses.

Wool
Characteristically very short, and shed annually.
Fibre diameter: 30–32 microns.
Staple length: 25–50 mm (1–2 inches).
Fleece weight: Range 1–1.8 kg (2.2–4 lb); Average 1.4 kg (3 lb).
Uses: Not used for commercial production.

Breeding/Lambing
190–210 percent.

Numbers
Wiltshire Horn: Under 300.
Poll Wiltshire: 2,000.

Useful Addresses

New Zealand Sheepbreeders' Association
Mr T.F. Burrows
P.O. Box 20-094
Christchurch

Cheviot Sheep Society of New Zealand
Miss Judith Pascoe
Creyke Road
RD1
Darfield
Canterbury

Coopworth Sheep Society of New Zealand
Mr C. Logan
P.O. Box 169
Lincoln University
Canterbury

New Zealand Wiltshire Sheep Breeders' Association
Mr J. Morrison
RD2
Marton

Perendale Sheep Society
Miss Judith Pascoe
Creyke Road
RD1
Darfield
Canterbury

Romney Sheep Breeders' Association
Miss C.H. Ramsay
P.O. Box 231
Feilding

Sheep Dairy Association of New Zealand
The Secretary
Okapua Homestead
RD3
Gore

Southdown Sheep Society of New Zealand
Miss C.H. Ramsay
P.O. Box 231
Feilding

Wools of New Zealand
Julie Everett-Hincks
Technical Officer
PO Box 3225
Wellington 1

Glossary

Apparel Clothing.
Braids Wool fibres that have become interwoven to produce a band.
Bulk A measure of wool's ability to fill space, measured in cubic centimetres per gram.
Carding A method of preparing wool by means of wire strips fixed to rollers.
Clip The wool that has been shorn from a sheep.
Cloth A woven material.
Coatings Materials used to make coats.
Commercial flock Sheep farmed for the wool and meat they produce. A flock may contain more than one breed of sheep, or various types of cross-bred sheep.
Crimp The folds in a wool fibre. They may be corkscrew-shaped (helical) or flat (planar). Helical crimp gives the wool more bulk.
Cross-breeding Mating a male and female of two different breeds. The resulting lambs have some of the characteristics of both parent breeds.
Dags (short for 'daglocks') Lumps of matted and soiled wool, found mainly under the tail and hind legs.
Dual-purpose sheep Sheep from which production has an equal emphasis on wool and meat.
Early maturing Reaching an ideal market condition (muscle, meat/fat ratio) at an earlier age than average. Early-maturing lambs do not grow any faster than late-maturing ones, so they are smaller and lighter in weight when marketed.
Ewe Female sheep.
Facial eczema A liver disease caused by a pasture fungus.
Felt A type of cloth made of wool, or wool and cotton, compacted together by rolling, beating and pressing with lees (liquor sediment) or size (a gelatinous solution).
Fibre diameter The diameter of an individual wool fibre. Measured in microns (one-thousandths of a millimetre).
Flannel A soft woollen material of open texture with a light nap (see Nap).
Fleece The wool covering an individual sheep.
Fleece rot See Wool rot.
Handle The way wool feels when handled; e.g. soft, harsh, or springy.
Hogget A sheep that has acquired its first pair of permanent teeth, usually about one year old. Males are ram hoggets, females are ewe hoggets.
Kemps Short, coarse hairy fibres that may be found in fleece, and will not take dye.
Lamb A young sheep that has not acquired its first pair of permanent teeth (which usually occurs at about one year of age).
Lambing percentage The average number of lambs produced per ewe over a flock or breed population, expressed as a percentage; e.g. one lamb per ewe = 100 percent; two lambs per ewe = 200 percent. It is usually calculated at docking time.
Lock A portion of fairly long wool that hangs together.
Lustre A shiny surface that is a feature of some wools.
Medullated fibre A fibre with an inner core of air cells which make it stiffer and more resistant to compression.

Micron One-thousandth of a millimetre. The diameter of wool fibre is measured in microns.
Mohair The fine hair from an Angora goat.
Nap The smooth or even surface produced on cloth or other fabric by cutting or smoothing the fibre or pile.
Open face A sheep's face not covered with wool.
Open fleece A fleece in which the wool fibres are not densely packed together, and which opens easily when handled.
Overcoatings Materials used to make overcoats.
Poll The area on top of a sheep's head, between and in front of the ears.
Polled Naturally having no horns.
Prime lambs Lambs specially produced for meat consumption at an early age.
Ram Male sheep.
Roller lapping A process that uses a roller to layer wool, or wool and cotton, to make felt.
Roman nose The bridge of a nose which has a pronounced convex curve.
Scouring The process by which wool is cleansed of grease and impurities.
Scurs Small or rudimentary horns found in some females of horned breeds.
Sire The male parent of a lamb — another term for ram.
Spring The ability of wool to regain shape after compression (see also Bulk, Medullated fibre).
Staple The wool fibres in a sheep's fleece. The term is also used to describe their character; e.g. fine or long staple.
Staple length The length of the wool fibres.
Stud flock Pedigree sheep of a specific breed, bred to maintain or improve the breed's standards.
Terminal sire A ram whose progeny are destined only for slaughter.
Textiles Fabrics made on a loom or a felting or knitting machine.
Tweed A twilled wool or wool-and-cotton fabric with an unfinished surface, used mainly for outer garments.
Twill A woven fabric in which the weft threads pass alternately over one warp thread and under two or more, producing diagonal ribs or lines.
Two-tooth A young sheep in which the second pair of permanent incisor teeth has erupted. These are normally visible between 16 and 22 months of age.
Upholstery Textile coverings for furniture.
Warp Threads stretched lengthwise in a loom, to be crossed by other threads (weft).
Web The material produced by the process of weaving.
Weft Cross-threads woven into warp to make material (web).
Wether A castrated male sheep.
Wool rot Rotting of wool fibres, more likely to occur in harsher fleeces into which rain can more easily penetrate and scour out the natural protective grease.
Worsted Woollen yarn, manufactured by mechanical combing that removes the shorter wool fibres and retains those that are longer.
Yarn A spun thread, especially of the kind prepared for weaving or knitting.

Bibliography

Ponting, Kenneth, *Sheep of the World*. Blandford Press, Dorset, UK, 1980.

Warman, Mike, *What Sheep is That?* GP Books, Wellington, 1991.

Wools of New Zealand, *New Zealand Sheep and their Wool*. Wools of New Zealand, Wellington, 1994 (5th ed.).

Perendale New Zealand, November 1995.